NANOTECHNOLOGY AS A NATIONAL SECURITY ISSUE

NANOTECHNOLOGY AS A NATIONAL SECURITY ISSUE

JOHN F. SARGENT

Nova Science Publishers, Inc.
New York

Copyright © 2009 by Nova Science Publishers, Inc.

All rights reserved. No part of this book may be reproduced, stored in a retrieval system or transmitted in any form or by any means: electronic, electrostatic, magnetic, tape, mechanical photocopying, recording or otherwise without the written permission of the Publisher.

For permission to use material from this book please contact us:
Telephone 631-231-7269; Fax 631-231-8175
Web Site: http://www.novapublishers.com

NOTICE TO THE READER

The Publisher has taken reasonable care in the preparation of this book, but makes no expressed or implied warranty of any kind and assumes no responsibility for any errors or omissions. No liability is assumed for incidental or consequential damages in connection with or arising out of information contained in this book. The Publisher shall not be liable for any special, consequential, or exemplary damages resulting, in whole or in part, from the readers' use of, or reliance upon, this material. Any parts of this book based on government reports are so indicated and copyright is claimed for those parts to the extent applicable to compilations of such works.

Independent verification should be sought for any data, advice or recommendations contained in this book. In addition, no responsibility is assumed by the publisher for any injury and/or damage to persons or property arising from any methods, products, instructions, ideas or otherwise contained in this publication.

This publication is designed to provide accurate and authoritative information with regard to the subject matter covered herein. It is sold with the clear understanding that the Publisher is not engaged in rendering legal or any other professional services. If legal or any other expert assistance is required, the services of a competent person should be sought. FROM A DECLARATION OF PARTICIPANTS JOINTLY ADOPTED BY A COMMITTEE OF THE AMERICAN BAR ASSOCIATION AND A COMMITTEE OF PUBLISHERS.

LIBRARY OF CONGRESS CATALOGING-IN-PUBLICATION DATA

Available upon request

ISBN 978-1-60692-070-1

Published by Nova Science Publishers, Inc. ✢ New York

CONTENTS

Preface vii

Chapter 1 Nanotechnology and U.S. Competitiveness: Issues and Options 1

Chapter 2 Nanotechnology: A Policy Primer 35

Index 51

PREFACE

As explained in Chapter 1, the projected economic and societal benefits of nanotechnology have propelled global investments by nations and companies. The United States launched the first national nanotechnology initiative in 2000. Since then, more than 60 nations have launched similar initiatives. In 2006, global public investment in nanotechnology was estimated to be $6.4 billion, with an additional $6.0 billion provided by the private sector. More than 600 nanotechnology products are now in the market, generally offering incremental improvements over existing products. However, proponents maintain that nanotechnology research and development currently underway could offer revolutionary applications with significant implications for the U.S. economy, national and homeland security, and societal well-being. These investments, coupled with nanotechnology's potential implications, have raised interest and concerns about the U.S. competitive position.

The data used to assess competitiveness in mature technologies and industries, such as revenues and market share, are not available for assessing nanotechnology. In fact, the U.S. government does not currently collect such data for nanotechnology, nor is comparable international data available. Without this information, an authoritative assessment of the U.S. competitive position is not possible. Alternatively, indicators of U.S. scientific and technological strength (e.g., public and private research investments, nanotechnology papers published in scientific journals, patents) may provide insight into the current U.S. position and serve as bellwethers of future competitiveness. By these criteria, the United States appears to be the overall global leader in nanotechnology. However, other nations are investing heavily and may lead in specific areas of nanotechnology. Some believe the U.S. leadership position in nanotechnology may not be as large as it has been in previous emerging technologies.

Efforts to develop and commercialize nanotechnology face a variety of challenges — e.g., technical hurdles; availability of capital; environmental, health, and safety concerns; and immature manufacturing technology and infrastructure. Some advocate a more active federal government role in overcoming these challenges, including funding to aid in the translation of research to commercial products; general and targeted tax provisions; incentives for capital formation; increased support for development of manufacturing and testing infrastructure, standards and nomenclature development, and education and training; creation of science, technology, and innovation parks; and efforts to establish a stable and predictable regulatory environment that keeps pace with innovation.

Some support a more limited federal role. Some who hold this view maintain that the market, free from government interventions, is most efficient. They assert that federal efforts can create market distortions and result in the federal government picking "winners and losers" among technologies, companies, and industries. Others oppose federal support for industrial research and applications, labeling such efforts "corporate welfare." Still others argue for a moratorium on nanotechnology R&D until environmental, health, and safety concerns are addressed.

As Chapter 2 presents, nanoscale science, engineering and technology — commonly referred to collectively as nanotechnology — is believed by many to offer extraordinary economic and societal benefits. Congress has demonstrated continuing support for nanotechnology and has directed its attention primarily to three topics that may affect the realization of this hoped for potential: federal research and development (R&D) in nanotechnology; U.S. competitiveness; and environmental, health, and safety (EHS) concerns. This report provides an overview of these topics — which are discussed in more detail in current and upcoming CRS reports — and two others: nanomanufacturing and public understanding of and attitudes toward nanotechnology.

The development of this emerging field has been fostered by significant and sustained public investments in nanotechnology R&D. Nanotechnology R&D is directed toward the understanding and control of matter at dimensions of roughly 1 to 100 nanometers. At this size, the properties of matter can differ in fundamental and potentially useful ways from the properties of individual atoms and molecules and of bulk matter. Since the launch of the National Nanotechnology Initiative (NNI) in 2000, Congress has appropriated approximately $8.4 billion for nanotechnology R&D. More than 60 nations have established similar programs. In 2006 alone, total global public R&D investments reached an estimated $6.4 billion, complemented by an estimated private sector investment of $6.0 billion. Data on economic outputs that are used to assess competitiveness in mature technologies and industries, such as revenues and market share, are not available for assessing nanotechnology. Alternatively, data on inputs (e.g., R&D expenditures) and non-financial outputs (e.g. scientific papers, patents) may provide insight into the current U.S. position and serve as bellwethers of future competitiveness. By these criteria, the United States appears to be the overall global leader in nanotechnology, though some believe the U.S. lead may not be as large as it has been for previous emerging technologies.

Some research has raised concerns about the safety of nanoscale materials. There is general agreement that more information on EHS implications is needed to protect the public and the environment; to assess and manage risks; and to create a regulatory environment that fosters prudent investment in nanotechnology-related innovation. Nanomanufacturing — the bridge between nanoscience and nanotechnology products — may require the development of new technologies, tools, instruments, measurement science, and standards to enable safe, effective, and affordable commercial-scale production of nanotechnology products. Public understanding and attitudes may also affect the environment for R&D, regulation, and market acceptance of products incorporating nanotechnology.

In 2003, Congress enacted the 21st Century Nanotechnology Research and Development Act providing a legislative foundation for some of the activities of the NNI, addressing concerns, establishing programs, assigning agency responsibilities, and setting authorization levels. Both the House of Representatives and the Senate remain actively engaged in the NNI, holding hearings in 2007 and 2008 related to possible amendments to, and reauthorization of, the act. Policy issues related to the NNI may be addressed in this process or through separate legislation.

Chapter 1

NANOTECHNOLOGY AND U.S. COMPETITIVENESS: ISSUES AND OPTIONS[*]

John F. Sargent
Science and Technology Policy Resources,
Science, and Industry Division

SUMMARY

The projected economic and societal benefits of nanotechnology have propelled global investments by nations and companies. The United States launched the first national nanotechnology initiative in 2000. Since then, more than 60 nations have launched similar initiatives. In 2006, global public investment in nanotechnology was estimated to be $6.4 billion, with an additional $6.0 billion provided by the private sector. More than 600 nanotechnology products are now in the market, generally offering incremental improvements over existing products. However, proponents maintain that nanotechnology research and development currently underway could offer revolutionary applications with significant implications for the U.S. economy, national and homeland security, and societal well-being. These investments, coupled with nanotechnology's potential implications, have raised interest and concerns about the U.S. competitive position.

The data used to assess competitiveness in mature technologies and industries, such as revenues and market share, are not available for assessing nanotechnology. In fact, the U.S. government does not currently collect such data for nanotechnology, nor is comparable international data available. Without this

---

[*] This report is excerpted from CRS report #106153, May 15, 2008

information, an authoritative assessment of the U.S. competitive position is not possible. Alternatively, indicators of U.S. scientific and technological strength (e.g., public and private research investments, nanotechnology papers published in scientific journals, patents) may provide insight into the current U.S. position and serve as bellwethers of future competitiveness. By these criteria, the United States appears to be the overall global leader in nanotechnology. However, other nations are investing heavily and may lead in specific areas of nanotechnology. Some believe the U.S. leadership position in nanotechnology may not be as large as it has been in previous emerging technologies.

Efforts to develop and commercialize nanotechnology face a variety of challenges — e.g., technical hurdles; availability of capital; environmental, health, and safety concerns; and immature manufacturing technology and infrastructure. Some advocate a more active federal government role in overcoming these challenges, including funding to aid in the translation of research to commercial products; general and targeted tax provisions; incentives for capital formation; increased support for development of manufacturing and testing infrastructure, standards and nomenclature development, and education and training; creation of science, technology, and innovation parks; and efforts to establish a stable and predictable regulatory environment that keeps pace with innovation.

Some support a more limited federal role. Some who hold this view maintain that the market, free from government interventions, is most efficient. They assert that federal efforts can create market distortions and result in the federal government picking "winners and losers" among technologies, companies, and industries. Others oppose federal support for industrial research and applications, labeling such efforts "corporate welfare." Still others argue for a moratorium on nanotechnology R&D until environmental, health, and safety concerns are addressed.

INTRODUCTION

Nanotechnology is believed by many to be one of the most promising areas of technological development and among the most likely to deliver substantial economic and societal benefits to the United States in the 21st century. With so much potentially at stake, a global competition has emerged among nations and companies to develop and capture the value of nanotechnology products.

Competitiveness generally refers to the comparative ability of a nation or company to bring products or services to markets. Assessments of competitive strength generally rely on indicators such as revenues, market share, and trade. However, since nanotechnology is still largely in an early stage of development the U.S. government does not collect this type of data for nanotechnology products. In addition, nanotechnology is not a discrete industry, but rather a technology applied across a wide range of products in disparate industries for which nanotechnology products generally account for a small fraction of total sales. For these reasons, an assessment of U.S. industrial competitiveness in nanotechnology — in the same manner that analysts would assess the competitiveness of mature industries — is not possible at this time. Alternatively, this report reviews national nanotechnology research and development (R&D) investments, scientific papers, and patents as indicators of current U.S. scientific and technological competitiveness and potential indicators of future industrial competitiveness in nanotechnology products.

The federal government has played a central role in catalyzing U.S. R&D efforts. In 2000, President Clinton launched the U.S. National Nanotechnology Initiative (NNI), the world's first integrated national effort focused on nanotechnology. The NNI has enjoyed strong, bipartisan support from the executive branch, the House of Representatives, and the Senate. Each year, the President has proposed increased funding for federal nanotechnology R&D, and each year Congress has provided additional funding. Since the inception of the NNI, Congress has appropriated a total of $8.4 billion for nanotechnology R&D intended to foster continued U.S. technological leadership and to support the technology's development, with the long-term goals of: creating high-wage jobs, economic growth, and wealth creation; addressing critical national needs; renewing U.S. manufacturing leadership; and improving health, the environment, and the overall quality of life.

The United States is not alone in seeking to tap the perceived potential of nanotechnology. Following the creation of the NNI, more than 60 nations have established their own national nanotechnology initiatives, many based on the U.S. model. Estimated global annual public investments in nanotechnology, including those of the United States, reached $6.4 billion in 2006, with another $6.0 billion invested by the private sector[1]. In addition, some countries have established strategic plans; nanotechnology-focused science, technology, and innovation parks; venture capital funds; and other policies and programs to accelerate the translation of nanotechnology research into products to exploit its economic potential. These investments and policies, coupled with generally optimistic expectations, have raised interest and concerns about the global competitive position of the United States in the development and commercialization of nanotechnology.

In 2003, Congress enacted the 21st Century Nanotechnology Research and Development Act (P.L. 108-153) assigning responsibilities and initiating research efforts to address key challenges. In the act, Congress explicitly established global

---

[1] Profiting From International Nanotechnology, Lux Research, December 2006. p. 2.

technological leadership, commercialization, and national competitiveness as central goals of the NNI:

> **National Nanotechnology Initiative**
>
> The National Nanotechnology Initiative is a federal government R&D initiative, coordinated by the White House, involving 25 departments and agencies, including 13 that conduct nanotechnology R&D.
>
> President Bill Clinton launched the NNI in 2000, and Congress provided $464 million in R&D funding to NNI agencies in FY2001. Since then, Congress has more than tripled annual funding, providing $1.49 billion in FY2008, bringing cumulative appropriations for NNI activities to $8.4 billion. President Bush has requested $1.53 billion for the NNI in FY2009.
>
> The NNI budget is an aggregation of the nanotechnology components of the individual budgets of NNI-participating agencies. The NNI budget is not a single, centralized source of funds that is allocated to individual agencies. In fact, agency nanotechnology budgets are developed internally as part of each agency's overall budget development process. These budgets are subjected to review, revision, and approval by the White House Office of Management and Budget and become part of the President's annual budget submission to Congress. The NNI budget is then calculated by aggregating the nanotechnology components of the appropriations provided by Congress to each federal agency.
>
> In 2003, Congress provided a statutory foundation for some of the activities of the NNI through the 21st Century Nanotechnology Research and Development Act (P.L. 108-153). The Act established a National Nanotechnology Program (NNP) and provided authorizations totaling $3.7 billion over four years (FY2005-FY2008) for five NNI agencies: the National Science Foundation, Department of Energy, NASA, National Institute of Standards and Technology, and the Environmental Protection Agency. In total, Congress appropriated $2.9 billion for these agencies during this period, or approximately 79% of total authorized funding. The Act did not address the participation of several agencies that fund nanotechnology R&D under the NNI, including the Department of Defense, National Institutes of Health, and the Department of Homeland Security.

The activities of the Program shall include —
...ensuring United States global leadership in the development and application of nanotechnology;
advancing the United States productivity and industrial competitiveness through stable, consistent, and coordinated investments in long-term scientific and engineering research in nanotechnology;
accelerating the deployment and application of nanotechnology research and development in the private sector, including start-up companies; ...and
encouraging research on nanotechnology advances that utilize existing processes and technologies.

Congress has expressed interest in understanding whether the current level of appropriations and the portfolio of activities pursued by the NNI is sufficient to achieve these goals. There are a variety of perspectives on the sufficiency and balance of activities and resources devoted to nanotechnology R&D, regulation, and infrastructure. This report provides an overview of nanotechnology, current and anticipated applications, indicators of U.S. scientific and technological strength, and issues and options Congress may opt to consider for the federal role, if any, in promoting the nation's competitive position in nanotechnology.

CURRENT AND ANTICIPATED APPLICATIONS

Nanotechnology — a term encompassing nanoscale science, technology, and engineering — involves the understanding and control of matter at scales between 1 and 100 nanometers. A nanometer is one-billionth of a meter; by way of comparison, the width of a human hair is approximately 80,000 nanometers.

At this size, the physical, chemical, and biological properties of materials can differ in fundamental and potentially useful ways from the properties of individual atoms and molecules, on the one hand, or bulk matter, on the other hand. Nanotechnology research and development is directed toward understanding and creating improved materials, devices, and systems that exploit these properties as they are discovered and characterized[2].

Most nanotechnology products currently on the market — such as faster computer processors, higher density memory devices, better baseball bats, lighter-weight auto parts, stain-resistant clothing, cosmetics, and clear sunscreen —are evolutionary in nature, offering valuable, but generally modest, economic and societal benefits.

---

[2] The National Nanotechnology Initiative Strategic Plan, Nanoscale Science, Engineering, and Technology Subcommittee, National Science and Technology Council, The White House, December 2007.

Over the next five to ten years, proponents see nanotechnology offering the potential for additional evolutionary improvements in existing products. Beyond the next ten years, they believe that nanotechnology could deliver revolutionary advances that could transform or replace existing products and industries, and create entirely new ones. Some hoped-for applications discussed by the technology's proponents, involving various degrees of speculation and varying time-frames, include: new prevention, detection, and treatment technologies that reduce death and suffering from cancer and other deadly diseases[3]; new organs to replace damaged or diseased ones;[4] clothing that protects against toxins and pathogens;[5] clean, inexpensive, renewable power through energy creation, storage, and transmission technologies;[6] universal access to safe water through portable, inexpensive water purification systems;[7] energy efficient, low-emission "green" manufacturing systems;[8] high-density memory systems capable of storing the entire Library of Congress collection on a device the size of a sugar cube;[9] agricultural technologies that increase yield and improve nutrition, reducing global hunger and malnutrition;[10] self-healing materials;[11] powerful, small, inexpensive sensors that can warn of minute levels of toxins and pathogens in air, soil, or water, and alert us to changes in the environment;[12] and environmental remediation of contaminated industrial sites.[13] Proponents in government, academia, and industry also maintain that nanotechnology could make substantial

---

[3] National Cancer Institute website. [http://nano.cancer.gov/resource_center/tech_backgrounder.asp]

[4] Ibid.

[5] Risbud, Aditi. "Fruit of the Nano Loom," Technology Review, February 2006.

[6] Nanoscience Research for Energy Needs, Nanoscale Science, Engineering, and Technology Subcommittee, National Science and Technology Council, The White House, December 2004.

[7] Risbud, Aditi. "Cheap Drinking Water from the Ocean," Technology Review, June 2006.

[8] Selko, Adrienne. "New Nanotechnology-Based Coatings are Energy Efficient and Environmentally Sound," Industry Week, August 22, 2007. "Tomorrow's Green Nanofactories," Science Daily, July 11, 2007.

[9] National Nanotechnology Initiative — Leading to the Next Industrial Revolution, Interagency Working Group on Nanoscience, Engineering, and Technology, National Science and Technology Council, The White House. [http://www.ostp.gov/NSTC/html/iwgn/iwgn.fy01budsuppl/nni.pdf]

[10] 21st Century Agriculture: A Critical Role for Science and Technology, U.S. Department of Agriculture, June 2003; Nanoscale Science and Engineering for Agriculture and Food Systems, draft report on the National Planning Workshop, submitted to the Cooperative State Research, Education, and Extension Service of the U.S. Department of Agriculture, July 2003.

[11] Nanotechnology in Space Exploration, Nanoscale Science, Engineering, and Technology Subcommittee, National Science and Technology Council, The White House, August 2004.

[12] Nanotechnology and the Environment, Nanoscale Science, Engineering, and Technology Subcommittee, National Science and Technology Council, The White House, May 2003.

[13] Proceedings of the U.S. Environmental Protection Agency Workshop on Nanotechnology for Site Remediation, U.S. Environmental Protection Agency, October 2005.

contributions to national defense, homeland security, and space exploration and commercialization.

Many areas of public policy could affect the ability of the United States to capture the future economic and societal benefits associated with these investments. Congress established programs, assigned responsibilities, authorized funding levels, and initiated research to address key issues in the 21st Century Nanotechnology Research and Development Act. The agency budget authorizations provided for in this act extend through FY2008 (see text box, "National Nanotechnology Initiative," for discussion of authorizations and appropriations).[14] Both the House and Senate have held committee hearings related to amending and reauthorizing this act in 2008. A companion report, CRS Report RL34401, *The National Nanotechnology Initiative: Overview, Reauthorization, and Appropriations Issues*, by John F. Sargent, provides an overview of nanotechnology; the history, goals, structure, and federal funding of the National Nanotechnology Initiative; and issues related to its management and reauthorization.

As the state of nanotechnology knowledge has advanced, new policy issues have emerged. In addition to providing funding for nanotechnology R&D, Congress has directed increased attention to issues affecting the U.S. competitive position in nanotechnology and related issues, including nanomanufacturing; commercialization; environmental, health, and safety concerns; workforce development; and international collaboration. Views and options related to these issues are presented later in this report.

U.S. COMPETITIVENESS INDICATORS

Nanotechnology is, by and large, still in its infancy. Accordingly, measures such as revenues, market share, and global trade statistics — indicators often used to assess and track U.S. competitiveness in other technologies and industries — are not available for assessing the U.S. position in nanotechnology. To date, the federal government does not collect data on nanotechnology-related revenues, trade or employment, nor is comparable international government data available.

---

[14] Under the act, Congress authorized $3.7 billion over four years (FY2005-FY2008) for five NNI agencies: the National Science Foundation, Department of Energy, NASA, National Institute of Standards and Technology, and the Environmental Protection Agency. In total, Congress appropriated $2.9 billion for these agencies during this period, or approximately 79% of authorized funding. Several NNI agencies — including two with large nanotechnology R&D budgets, the Department of Defense and National Institutes of Health — did not receive budget authorizations under the act.

Nevertheless, many experts believe that the United States is the global leader in nanotechnology. For example, a survey of U.S. business leaders in the field of nanotechnology showed 63% believe that the United States is leading other countries in nanotechnology R&D and commercialization while only 7% identified the United States as lagging behind other countries.[15]

However, some believe that in contrast to many previous emerging technologies — such as semiconductors, satellites, software, and biotechnology — the U.S. lead appears narrower and the investment level, scientific and industrial infrastructure, technical capabilities, and science and engineering workforces of other nations are more substantial than in the past. Charles Vest, president of the National Academies of Engineering and a member of the President's Council of Advisors on Science and Technology (PCAST), asserted early in 2008 that nanotechnology was the first emerging technology "where we [the United States] don't have a huge lead." Vest added that it was also the first emerging technology in which the federal government's efforts included "commercialization as a specific goal" and thus was "the first real test" of the United States' "loosely-coupled public-private partnership in the new competitive environment."[16]

Evidence of commercialization of nanotechnology-based products is generally available. For example, the Woodrow Wilson International Center for Scholars' Project on Emerging Nanotechnologies counts more than 600 company-identified nanotechnology products on the market, more than half of which are produced by companies based in the United States.[17] Some private organizations have attempted to estimate current nanotechnology-derived revenues and to estimate future revenues. For example, Lux Research estimates that products incorporating nanotechnology produced $50 billion in global revenues in 2006[18] (less than 0.1% of global manufacturing output), and that by 2014 revenues will reach $2.6 trillion or 15% of projected global manufacturing output.[19]

---

[15] "Survey of U.S. Nanotechnology Executives," conducted by Small Times Magazine and the Center for Economic and Civic Opinion at the University of Massachusetts-Lowell, Fall 2006. [http://www.masseconomy.org/pdfs/ nano_survey_report_gocefd2.pdf]

[16] Personal notes from PCAST meeting held January 8, 2008.

[17] "Consumers Talk Nano," press release, Project on Emerging Nanotechnologies, Woodrow Wilson International Center for Scholars, October 22, 2007. [http://www.wilsoncenter.org/ index.cfm?topic_id=166192& fuseaction=topics.item&news_id=297072]

[18] The Nanotech Report, 5th Edition, Vol. 1, Lux Research, November 2007. p. iii.

[19] Nordan, Matthew, president, LuxResearch, Inc. Testimony before the Subcommittee on Science, Technology, and Innovation, Committee on Commerce, Science and Transportation, U.S. Senate. Hearing on "National Nanotechnology Initiative: Charting the Course for Reauthorization." 110th Cong., 2nd Sess., April 24, 2008.

In the absence of comprehensive and authoritative economic output data (e.g., revenues, market share, trade), indicators such as inputs (e.g., public and private research investments) and non-financial outputs (e.g., scientific papers, patents) are now used to gauge a nation's current and future competitive position in emerging technologies. These indicators offer insights into nations' scientific and technological strength which may serve as a foundation for future product and process innovation.

However, research and development investments, scientific papers, and patents may not provide reliable indicators of the United States' current or future competitive position. Scientific and technological leadership may not necessarily result in commercial leadership and/or in national competitiveness for the following reasons:

Basic research in nanotechnology may not translate into viable commercial applications. Though no formal assessment of the composition of the NNI budget has been made, there is general consensus that the NNI investment since its inception has been focused on basic research. The National Science Foundation defines the objective of basic research as seeking "to gain more comprehensive knowledge or understanding of the subject under study without applications in mind."[20] Therefore, while basic research may underpin applied research, development, and commercialization, that is not its primary focus or intent. In general, basic research can take decades[21] to result in commercial applications, and many advances in scientific understanding may not present commercial opportunities.

Basic research is generally available to all competitors. Even when basic research presents the potential for commercial exploitation, it may not deliver national advantage. Open publication and free exchange of research results are guiding principles of federally funded fundamental research[22] and research

---

[http://commerce.senate.gov/public/_files/LuxResearchSenateCommerceCommitteetestimony4242008.pdf]

[20] Science and Engineering Indicators 2008, National Science Foundation, January 2008.

[21] For example, the first working fuel cell was built in 1843, but the first semi-commercial use of a fuel cell did not occur for more than a hundred years when the technology was used in NASA's Project Gemini space program. Even today, commercial production and use of fuel cells is limited and federal technology development efforts continue.

[22] National Security Decision Directive 189 states that, "It is the policy of this Administration that, to the maximum extent possible, the products of fundamental research remain unrestricted." The directive allows for restriction of some results through national security classification. Fundamental research is defined in the directive as including both basic and applied research in science and engineering, but distinct from proprietary research and industrial development, design, production, and product utilization. For further information on U.S. policy toward unrestricted access to federally-funded fundamental research, see CRS Report RL31695,

conducted by U.S. colleges and universities. This approach may allow for the rapid expansion of global scientific and technical knowledge as new work is built on the scaffolding of previous work. However, the information is available to all competitors, U.S. and foreign alike, and thus may not confer competitive advantage to the United States.

U.S.-based companies may conduct production and other work outside of the United States. In today's economy, supply chains are global and the work required to develop, design, produce, market, sell, and service products is generally conducted where it can be done most efficiently. Even if U.S.-based companies successfully develop and bring nanotechnology materials and products to market, work may be conducted, and the economic value captured, outside of the United States. Federal policies and investments may offer tools that can make the United States the most attractive place for companies to conduct a greater share of value-adding activities, contributing to U.S. economic growth and job creation.

U.S.-educated foreign students may return home to conduct research and create new businesses. In the era following World War II, many of the most gifted and talented students from around the world were attracted to the science and engineering programs of U.S. colleges and universities. For many years, many of those who graduated from these programs decided to stay in the United States and contributed to U.S. global scientific, engineering, and economic leadership. Today, many foreign students educated in the United States have economic opportunities in their home countries that did not exist for previous generations. Some nations are making strong appeals and offering significant incentives for their students to return home to conduct research and create enterprises. Thus, federal support for universities, in general, and scientific and engineering research activities, in particular, may contribute to the development of leading scientists and engineers who might return to their home countries to exploit the knowledge, capabilities, and networks developed in the United States.

Small businesses may lack the resources needed to bring their nanotechnology innovations to market. Federal programs, such as the Small Business Innovation Research (SBIR) program and the Small Business Technology Transfer (STTR) program, support leading-edge nanotechnology research by small innovative firms. Federally funded university research can produce small start-up ventures. These small businesses may develop commercially valuable technology, and even successfully develop new

---

Balancing Scientific Publication and National Security Concerns: Issues for Congress, by Dana A. Shea.

nanotechnology materials, tools, processes, or products, but lack the capital, infrastructure, or sales and distribution channels to effectively bring such advances to market.

U.S. companies with leading-edge, nanotechnology capabilities and/or their intellectual property may be acquired by foreign competitors. Foreign companies may acquire leading-edge nanotechnology companies or their intellectual property. This can take place, for example, as the result of an intentional business strategy to be acquired (a common exit strategy for start-up companies), a hostile takeover if the enterprise is a public company, or when a business has failed or is failing. In the latter case, the company or its intellectual property might be acquired at a fraction of its development cost or potential value.

U.S. policies or other factors may impede nanotechnology commercialization, make it unaffordable, or make it less attractive than foreign alternatives. Federal, state, and local policies (e.g., taxes, environmental and health regulations, ethical restrictions) and other factors (e.g., availability, quality, and cost of labor; proximity to markets; customer requirements; manufacturing infrastructure; public attitudes) may prevent or discourage commercialization of nanotechnology innovations in the United States. Companies may be prohibited from producing a commercially viable product in the United States, may be unable to do so affordably, or may find comparatively favorable conditions (e.g. lower taxes or tax holidays; fewer regulatory restrictions; qualified, available, and less costly workforce) outside the United States.

Comparisons of aggregate national data may be misleading. For example, a small nation with limited resources may be unable to pursue leading-edge research across a broad spectrum of nanotechnology-related disciplines and applications, and instead opt to seek technological dominance in a discreet area by investing in a limited set of disciplines and applications (or even a single one). In such a case, that country may become the strongest competitor in a given area, while analysis of aggregate numbers might obscure this strength. Alternatively, a rapidly developing nation may invest substantial capital in nanotechnology research, but lack key elements — such as a strong scientific and technological infrastructure; mature industry, service, and private capital infrastructure; experienced scientists, engineers, managers, and entrepreneurs; and/or a market-oriented business climate — needed to fully capitalize on such an investment.

In addition, the concept of a national competitive position may differ from the past as a result of increased globalization of research, technical talent, and production. For example, the world's leading-edge research in a field of nanotechnology might be conducted at an American university, by Chinese students, supported by research funds from a German-based corporation, with engineering underway in Russia, plans to manufacture in Taiwan, shipping by Greek-flagged vessels, and technical support provided online and by telephone from India. In such an example, the global distribution of knowledge workers, investors, and producers make the determination of national competitiveness more difficult.

Just as other countries might benefit from U.S. nanotechnology R&D, so too might the United States benefit from nanotechnology R&D conducted in other nations through a variety of means including studying published research results, acquiring or licensing patents, conducting joint business ventures, and by fostering a business environment that attracts production and related activities. Some economists assert that international R&D collaboration can benefit the United States as well by improving the productivity of the R&D process.[23]

With these caveats, the following section reviews input and non-economic output measures as indicators of the U.S. competitive position in nanotechnology.

Research and Development Investments[24]

National research and development investment is an input measure that may provide some perspective on how successful a nation and the firms within the nation may become in producing scientific and technical knowledge that can lead to innovative products and processes. However, the long-term value of these investments may be affected by a variety of factors such as: the capability of the scientists and engineers conducting the R&D and the tools available to them; the efficiency of the system (e.g., businesses, supply chains, infrastructure, innovation climate, government policies) for translating R&D results into commercial

---

[23] Economic Report of the President, Council of Economic Advisors, The White House, 1989. p. 225.
[24] Comparisons and aggregations of investments in R&D across organizational and national boundaries are fraught with imprecision and inaccuracy. One challenge with quantifying public or private investment in nanotechnology R&D relates to the definition of nanotechnology used by different governments and institutions. There is significant debate in and among federal agencies with respect to what should be considered (and thus counted) as nanotechnology R&D, not withstanding the NNI definition, as well as within and between companies, industries, and governments. In addition to substantive definitional disagreements about nanotechnology, some seek to take advantage of the cachet of the term "nano" or "nanotechnology." Strong interest in nanotechnology on the part of policymakers, investors, and consumers may induce some to characterize non-nanotechnology activities or products as nanotechnology, or to characterize an entire effort as nanotechnology R&D when in fact only a portion of it is devoted to nanotechnology. Conversely, some firms may not characterize nanotechnology research efforts as "nanotechnology" out of concern for potential negative reactions from customers or unwanted regulatory attention.

products; the fields of nanoscience and nanotechnology pursued; the balance in fundamental research, applied research, and development efforts; and balance in R&D directed at exploiting commercial opportunities, meeting societal needs (e.g., health, environment), addressing government missions (e.g., defense, homeland security), and non-directed efforts to expand the scientific knowledge frontier.

Public Investments

The United States has led, and continues to lead, all nations in public investments in nanotechnology R&D. However the estimated U.S. share of global public R&D investments in nanotechnology has fallen as other nations have established similar programs and increased funding. In the early part of this decade, many nations followed the U.S. lead and established formal national nanotechnology programs in recognition of the potential contributions nanotechnology may offer for economic growth, job creation, energy production and energy efficiency, environmental protection, public health and safety, and national security. According to Mike Roco, past chair of the National Science and Technology Council's (NSTC) Nanoscale Science, Engineering, and Technology (NSET) subcommittee, at least 60 countries have adopted national nanotechnology projects or programs.[25] Japan, Germany, and South Korea are making substantial sustained investments across a broad range of nanoscale science, engineering, and technology and are strong competitors for global leadership. More recently, China and Russia have increased investments in nanotechnology. In addition, others —such as Israel, Singapore, and Taiwan — have focused their resources on either a specific nanotechnology niche or on technology development (in contrast to fundamental research).[26]

Lux Research estimates that total 2006 public global R&D investments increased 10% over the 2005 level, reaching $6.4 billion. International investment levels can be compared using differing methods, producing substantially different perspectives on leadership. For example, using a currency exchange rate comparison, the United States ranks ahead of all others, with federal and state investments of $1.78 billion in 2006 (27.8% of global public R&D investments), followed by Japan ($975 million, 15.2%) and Germany ($563 million, 8.8%). When national investments are adjusted using purchasing power parity (PPP) exchange rates (which seek to equalize the purchasing power of currencies in

---

[25] Roco, M.C. "International Perspective on Government Nanotechnology Funding in 2005," Journal of Nanoparticle Research, 2005, Vol. 7(6).
[26] Ranking the Nations: Nanotech's Shifting Global Leaders, Lux Research, November 2005. p. 2.

different countries for a given basket of goods and/or services),[27] China ranks second in public nanotechnology spending in 2006 at $906 million, behind only the United States; Japan drops to third as its PPP-adjusted investment falls to $889 million.[28] Comparative international public funding for nanotechnology R&D is provided in Table 1.

Table 1. Top Ten Countries in Public Nanotechnology R&D, 2006

in millions of U.S. dollars using currency exchange rates		in millions of U.S. dollars using PPP exchange rates	
United States	1,775	United States	1,775
Japan	975	China	906
Germany	563	Japan	889
France	473	South Korea	563
South Korea	464	Germany	508
United Kingdom	280	France	403
China	220	Taiwan	249
Taiwan	132	United Kingdom	227
Russia	106	India	186
Canada	61	Russia	184

Source: Profiting from International Nanotechnology, Lux Research, December 2006.

Private Sector Investments

Private investments in nanotechnology R&D come from two primary sources, corporations and venture capital investors. Globally, corporations invested an estimated $5.3 billion in nanotechnology research and development in 2006. This figure represents a 19% increase over the 2005 estimate, a growth rate nearly twice that of global public R&D investments.[29] Faster growth in corporate R&D

---

[27] The use of PPP-adjusted numbers may distort the comparative value of national R&D investments since the "basket of goods" used to adjust prices generally (which may include food and other consumer items) may bear little resemblance to the goods and services purchased for nanotechnology research and development. In addition, while salaries of researchers in countries such as China may be lower than in other countries (i.e., more research can be bought with a dollar in China than in the United States), the cost of nanotechnology research equipment is likely to be essentially the same in all countries.

[28] Profiting from International Nanotechnology, Lux Research, December 2006. pp. 8-9.

[29] Profiting from International Nanotechnology, Lux Research, December 2006. pp. 25-26.

may be an indicator that nanotechnology research is moving closer to commercial production.

As with public R&D investments, on a PPP comparison basis, the United States led the world in 2006 in private sector R&D investments in nanotechnology with an estimated $1.9 billion investment, led by companies such as Hewlett-Packard, Intel, DuPont, General Electric, and IBM. Japan's $1.7 billion in private investments in nanotechnology R&D — led by companies such as Mitsubishi, NEC, and Hitachi —ranks a close second behind the United States. The private investments of companies headquartered in these two nations account for nearly three-fourths of corporate investment in nanotechnology R&D in 2006. In contrast to its high PPP ranking in public R&D investment, China ranks fifth in corporate investment, accounting for only about 3% of global private R&D investments in nanotechnology.[30]

Strength in an existing industry base may be a driver for private investment in nanotechnology innovations. For example, multi-walled carbon nanotubes (MCWNTs) offer significant improvements in lithium-ion (Li-ion) battery life. Japan's strength in Li-ion batteries is seen as a driving force in Japan's leading position in the manufacture of MWCNTs and Japanese companies' investments in ton-scale production capabilities.[31]

Venture capital investment — early-stage equity investment, generally characterized by high risk and high returns — provides another possible indicator of international competitiveness. In 2007, venture capital for nanotechnology reached an estimated $702 million worldwide of which U.S.-based companies received $632 million (approximately 90%).[32]

Scientific Papers

The quantity of peer-reviewed scientific papers published by scientists and engineers of each nation is one indicator of the scientific leadership of that nation. The scientific journals used to generate such counts tend to be considered among the most reliable and prestigious in the fields. Nevertheless, as a tool for assessing national competitiveness, this indicator has shortcomings. For example, paper counts do not assess the level or significance of contributions made by each of the authors. While an article may list a principal investigator, such as a university

---

[30] Profiting from International Nanotechnology, Lux Research, December 2006. pp. 9-10.
[31] International Assessment of R&D in Carbon Nanotube Manufacturing and Applications, World Technology Evaluation Center, Inc., June 2007.
[32] Private communication between CRS and Lux Research, Inc., April 28, 2008.

professor, as the lead author, the other authors, possibly graduate students or post-grads from a country other than that of the lead author, may have made the most important contributions to the work. Publication of a scientific paper may also represent a recognition of its unique scientific insights, yet offer little or no potential for useful applications or commercial relevance.

Output of Peer-Reviewed Papers

The United States leads all other nations in peer-reviewed nanotechnology papers published in scientific journals. A National Bureau of Economic Research (NBER) analysis reported that the United States' 24% share of global publication output was more than double that of the next most prolific nation, China.[33] However, this share represents a decline from the early 1990s when the United States accounted for approximately 40% of nanotechnology papers. The NBER working paper concludes, "Taken as a whole these data confirm that the strength and depth of the American science base points to the United States being the dominant player in nanotechnology for some time to come, while the United States also faces significant and increasing international competition."[34]

A quantitative analysis of published scientific papers comparing the United States to the Europe Union (EU) nations as a whole was prepared by United Kingdom-based Evaluametrics, Ltd. following an inquiry from the Congressional Research Service in December 2007.[35] Evaluametrics' analysis shows that the number of nanotechnology papers more than doubled between 2000 and 2005. Using a fractional count of papers,[36] the United States maintained about a 22% share of papers from 2000 to 2005. The EU27's[37] share of papers fell from 32% to 25% during this period, while China's share rose from 11% to 20%. Viewed from this perspective, the EU27 led the United States in output of nanotechnology-

---

[33] Zucker, L.G. and M.R. Darby. "Socio-Economic Impact of Nanoscale Science: Initial Results and Nanobank," National Bureau of Economic Research, March 2005. [http://www.nber.org/papers/w11181]

[34] Ibid.

[35] Nanotechnology Research Outputs 2000-2007: Interpretation of Results, prepared for CRS by Evaluametrics, Ltd. in December 2007. This analysis was performed using information from Thompson Scientific's Science Citation Index, selecting papers based on two filters, a list of specialist journals and a list of key nano-related words in the title of the paper.

[36] Using a fractional count approach, if a paper has multiple authors of different nations, a fraction of the paper is assigned to each country in proportion to the nations listed in the authors' addresses. Thus, a paper with three authors, two of which list U.S. addresses and one of which has an address in the United Kingdom, would be allocated as 0.67 (two-thirds) of a paper to the United States and 0.33 (one-third) of a paper to the United Kingdom.

[37] The EU27 represents the combined output for the 27 nations of the European Union.

related scientific papers, but the EU27 share has been in decline. China's share is approaching that of both the United States and the EU27 (see Figure 1).[38]

Using an integer count, with each paper assigned to the nation of the lead author's address, yields similar results. By this method, the EU27 led the world in 2006 with approximately 29% of all papers, followed by the United States with 25%, and China with approximately 23%. Evaluametrics' analysis of preliminary data shows that China may have surpassed the United States in share of papers in 2007.[39]

Evaluametrics' analysis of the papers by scientific disciplines reveals regional differences. The United States' articles were more heavily weighted toward the biological and medical fields, China's toward chemistry and engineering, and the EU27's toward the biological and medical fields, similar to the United States, but with a greater emphasis on physics and less on chemistry.

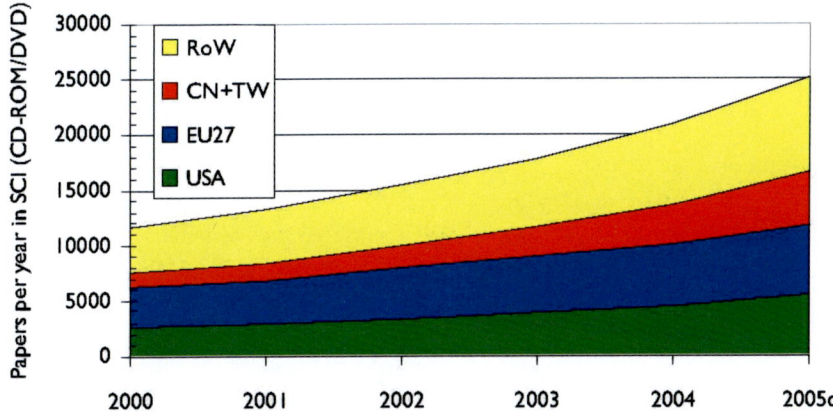

Source: Evaluametrics, Ltd., December 2007.
Note: RoW = Rest of World, CN+TW = China, including Taiwan, EU27 = nations of the European Union.

Figure 1. Nanotechnology Papers in the Science Citation Index, by Region, 2000-2005.

Evaluametrics also calculated for each country/region the share that nanotechnology papers represented as a percentage of all scientific papers. Dividing this percentage by the average for all nations yields a ratio the author calls a nation's "relative commitment" to nanotechnology. Of the 10 countries

---

[38] For purposes of this analysis, Evaluametrics attributes Taiwan's data to China.
[39] Nanotechnology Research Outputs 2000-2007: Interpretation of Results, Evaluametrics, Ltd., December 2007.

examined, South Korea, China, and Japan showed the highest relative commitment, while the United States and EU27 fell somewhat short of the world average (see Figure 2).

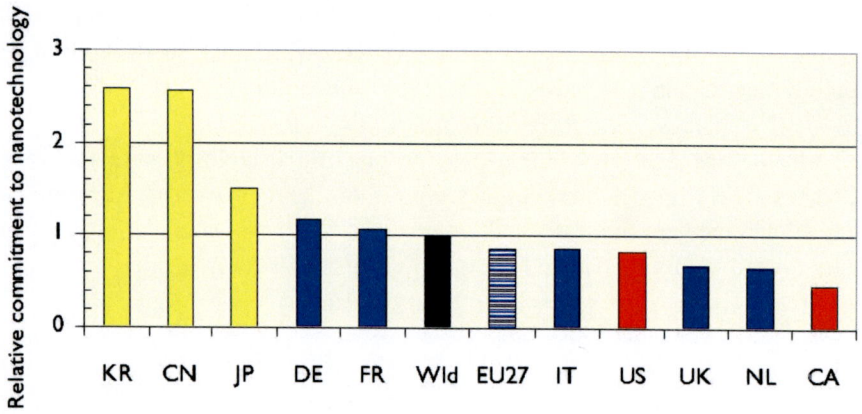

Source: Evaluametrics, Ltd., December 2007.
Note: KR = South Korea, CN = China (including Taiwan), JP = Japan, DE = Germany, FR = France, Wld = World, EU27 = nations of the European Union, IT = Italy, UK = United Kingdom, NL = The Netherlands, CA = Canada.

Figure 2. Relative Commitment of Ten Countries to Nanotechnology, 2003.

Citations to Peer-Reviewed Papers

Another measure of global leadership in nanotechnology is the quality and value of peer-reviewed papers. One measure of the quality and value of a paper is the frequency with which it is cited in other peer-reviewed papers. Evaluametrics' analysis shows that papers attributed to the United States are much more frequently cited than those attributed to China, the EU27, and the rest of the world as a whole. This held true overall and separately in each of the four disciplines examined (biology, chemistry, engineering, and physics).[40] The U.S. lead was particularly pronounced in biology. China fell below the world average number of citations in each of the four disciplines, as well as overall. The EU27 performed near the world average in engineering and physics, and somewhat higher in

---

[40] Using the Potential Citation Index metric based on the mean five-year citation counts to papers in the given journals.

chemistry. Using a slightly different citation metric,[41] 56% of U.S. papers have 10 or more citations, in contrast to only 38% for the EU27 and approximately 30% for China. The Netherlands and Germany lead the EU27 in papers with 10 or more citations with approximately 45% each.

Papers in "High-Impact" Journals

A second measure of the quality of a nation's papers is the share of its papers in influential journals. A review of *Science*, *Nature*, and *Physical Review Letters* (which PCAST refers to as "high impact journals") shows the United States accounted for more than 50% of nanotechnology-related papers in these journals in 2004. However, while the absolute number of papers in these journals attributed to the United States has grown continuously since 1991, the U.S. share of papers has fallen as other countries collectively increased their production at an even faster rate.[42]

Patents

Patent counts — assessments of how many patents are issued to individuals or institutions of a particular country — are another indicator used to assess a nation's competitive position. According to the U.S. Patent and Trade Office (USPTO), a patent grants ownership rights to a person who "invents or discovers any new and useful process, machine, manufacture, or composition of matter, or any new and useful improvement thereof." By this definition, patents may be an indicator of future value and national strength in a technology, product, or industry.

By this measure the United States position appears to be very strong. United States assignees dominate all other countries in patents issued by the USPTO. According to an analysis by the USPTO of patents in the United States and in other nations, U.S. origin inventors and assignees/owners have:

---

[41] Using the Actual Citation Index metric for papers published in 2003 and cited between 2003 and 2007.
[42] The National Nanotechnology Initiative at Five Years: Assessment and Recommendations of the National Nanotechnology Advisory Panel, President's Council of Advisors on Science and Technology, May 2005. [http://www.nano.gov/FINAL_PCAST_NANO_REPORT.pdf]

- the most nanotechnology-related U.S. patents by a wide margin;
- the most nanotechnology-related patent publications globally, but by a narrower margin (followed closely by Japan); and
- the most nanotechnology-related inventions that have patent publications in three or more countries, 31.7% — an indication of a more aggressive pursuit of international intellectual property protection and, by inference, of its perceived potential value. By this measurement, the United States is followed by Japan (26.9%), Germany (11.3%), Korea (6.6%), and France (3.6%).[43]

There has been rapid growth in nanotechnology patents in the USPTO and European Patent Office (EPO) patent databases. A 2007 study reports that the number of U.S. nanotechnology patents in the USPTO and EPO databases grew at a near exponential pace between 1980 and 2004. The study showed that each year since 1990, U.S. assignees have accounted for approximate two-thirds of all patents in the USPTO database. In 2004, U.S. assignees accounted for 66.9% of USPTO nanotechnology patents.[44]

An earlier study of USPTO data, covering patents from 1976 to 2002, also indicated U.S. nanotechnology patent leadership, with the United States accounting for more than 67% of patents (based on a full text search of patents for nanotechnology-related key words), followed by Japan, Germany, France, and Canada. In 2003, the United States, Japan, Germany, Canada, and France continued to rank in the top five, with the Republic of Korea and the Netherlands jumping several spots each to sixth and seventh places, respectively; Ireland and China made their first appearance in the top 20 nations. With respect to patent citations by subsequent patents — a possible indicator of the usefulness of a patent — the study showed U.S. nanotechnology patents dominated citations and reflected strong interactions with Japanese and German patents.[45]

Patent counts, however, have shortcomings in assessing future competitiveness. Experience shows that not all patents have equal value. Some

---

[43] Eloshway, Charles. "Nanotechnology Related Issues at the U.S. Patent and Trademark Office," Workshop on IPR in Nanotechnology: Lessons from Experiences Worldwide, Brussels, Belgium, April 2007. [ftp://ftp.cordis.europa.eu/pub/nanotechnology/docs/iprworkshop_eloshway_en.pdf]

[44] Li, Xin; Lin, Yiling; Chen, Hsinchun; Roco, Mihail C. "Worldwide Nanotechnology Development: A Comparative Study of USPTO, EPO, and JPO Patents (1976-2004)," Journal of Nanoparticle Research, Vol. 9, December 2007.

[45] Huang, Zan, Hsinchun Chen, and Zhi-kai Chen. "International Nanotechnology Development in 2003: Country, Institution, and Technology Field Analysis Based on USPTO Patent Database," Journal of Nanoparticle Research, Vol. 6, August 2004.

patents are "blockbusters" that largely define products and industries, and result in substantial wealth creation and competitive advantage for its owners. Other patents are useful, but offer only moderate value. And some patents are never realized in materials, products or processes. Patent counts do not attempt to assess the relative value of each patent, instead assigning equal value to each. Thus, a nation with a high patent count in nanotechnology may not benefit as much as a nation with fewer, but more valuable, patents.

In addition, companies may choose not to patent a particular idea — even one with significant value — for a variety of reasons. For example, public exposure of the idea as required by the patent application process may enable other companies to engineer around their patent, finding a way to do what the original patent accomplished, but in a way that is sufficiently different that it qualifies for a new patent. Alternatively, a company may be able to block the original patent holder from further improvements by filing patent applications that essentially block out potential improvements. Or, unscrupulous producers may ignore the patent and use the intellectual property without compensating the patent holder. Instead of filing for a patent, a company may choose instead to hold a valuable intellectual property as a trade secret. In contrast to patents which provide protection for defined periods, trade secrets can extend indefinitely. The Coca-Cola Company has held its formula(s) for Coke as a trade secret for over 100 years.[46]

THE FEDERAL ROLE IN U.S. COMPETITIVENESS: ISSUES AND OPTIONS

Congress has provided $8.4 billion in funding for nanotechnology R&D under the National Nanotechnology Initiative, and more than tripled annual funding since its inception. At the same time, other nations have established and bolstered their nanotechnology investments, programs, and policies, stirring debate about how the federal government can best ensure U.S. competitiveness in this field.

While there is broad consensus that U.S. competitiveness in nanotechnology is important, there are a wide variety of views about the role the federal government could or should play in supporting this objective. These perspectives are reflective of long-running policy debates over the appropriate role of government in promoting research, development, innovation, and industrial

---

[46] The Coca-Cola Company website. [http://www.thecoca-colacompany.com/heritage/worldcocacola.html]

competitiveness. Some argue that the federal role should be limited to funding basic research, R&D needed to meet agency mission requirements, and development of the U.S. scientific and technical workforce.[47] Some believe the federal government should also fund development efforts that move nanotechnology closer to commercial products, especially in light of its potential economic and societal benefits.[48] Others assert that the federal government's role in competitiveness should be limited to establishing a healthy business environment and allowing market forces and the private sector to shape U.S. competitiveness.[49] Among those concerned about nanotechnology's potential adverse implications for health, safety, and the environment are those who support a more active federal role in funding environmental, health, and safety (EHS) research to better understand, characterize, and regulate nanotechnology,[50] and those who prefer the federal government act to slow the development and commercialization of nanotechnology pending further EHS research.[51]

> **Survey of Industry Views on the Federal Role**
>
> A survey of U.S. nanotechnology business leaders indicated this community was divided on the desired level of government involvement in the development of nanomanufacturing technologies with 45% wanting "government to take the lead in R&D and commercialization incentives" and 43% wanting "limited participation." Another 11% of respondents said they wanted government to "stay out of it."
>
> Among the most significant barriers to growth identified by U.S. nanotechnology business leaders in a survey conducted by Small Times magazine and the University of Massachusetts-Lowell were: intellectual property issues (46%), lack of financing (45%), lack of available prototype facilities (43%), and lack of nanotechnology safety standards (36%). Ninety-two percent of respondents identified access to unique equipment and facilities as very important, and 91% identified access to

---

[47] Vannevar Bush's 1945 seminal report to President Truman, Science: The Endless Frontier, advocated this approach to support both government mission needs and industrial requirements.

[48] This perspective is exemplified by President Clinton's first technology policy statement, Technology for America's Economic Growth, A New Direction to Build Economic Strength, February 22, 1993. The Brookings Institution also advocates this position.

[49] Organizations such as the Cato Institute and the Heritage Foundation generally support such an approach.

[50] This position is held by many organizations, including the Woodrow Wilson International Center for Scholar's Project on Emerging Nanotechnologies, Environmental Defense, and DuPont.

[51] The ETC Group and Natural Resources Defense Council are advocates for this approach.

processes and tools to reduce time-to-market from R&D as very important.

Nearly three of five respondents indicated that they use or planned to use shared-use facilities at local universities, with science and engineering labs (25%), electronic labs (24%), and biotech labs (17%) topping the list, followed by specific diagnostic equipment (14%) and microfabrication labs (12%). More than three-fourths of the nanotechnology executives surveyed identified internal R&D as the primary source of expertise for the development of products and processes. Another 9% of executives identified industry associations or consortiums as their primary source of expertise, while only 7% identified collaboration with universities.

Source: "Survey of U.S. Nanotechnology Executives," conducted by Small Times Magazine and the Center for Economic and Civic Opinion at the University of Massachusetts-Lowell, Fall 2006.

Technology Development

Much of the public dialogue on how the government can advance U.S. strength in nanotechnology has focused on federal technology funding. Advocates for increased federal support put forth a variety of arguments.

Some believe that the federal government should provide increased funding for "downstream research," i.e., applied research and development closer to commercial products, including production prototypes.[52] Those who advocate this position generally assert that many promising research breakthroughs and early technology developments fail to make it to market. This failure, they argue, results from inadequate funding mechanisms to bring the technology to a state of maturity in which private corporations and other sources of capital are willing to invest in the technology — or in the company that holds the technology — to bring it to market. For example, they assert that investor demand for short-term returns can result in companies being unable to invest in higher-risk, longer-term technology development projects needed to sustain their viability in the future. Similarly, according to these advocates, venture capitalists and other investors often have exit strategies and/or seek returns in a timeframe (generally three to five years) inconsistent with the longer-term development horizons of emerging

---

[52] This position is held by many small technology businesses and start-ups.

and enabling technologies. With federal investments, say supporters, technical risk could be reduced to a level that enables promising research and early-stage technologies to overcome "the valley of death"[53] and reach the marketplace where the nation would be able to capture their economic and societal benefits.

One rationale offered by some economists for federal funding of R&D is that the private sector underinvests in R&D because many of the benefits (economic and societal) are captured by others, particularly where the results cannot be easily appropriated for production and profit.[54] However, as research moves closer to commercialization, private sector incentives to invest increase and the rationale for federal R&D funding is diminished.

Another argument for government R&D funding put forth by economists and others is that the development of emerging and enabling technologies may be beyond the ability of any single company or industry to develop due to high cost, high risk, lack of requisite expertise, and an inability to capture adequate returns. While a single company or industry may not be able to achieve adequate returns across its limited line of products and services, the economic benefits that accrue across all industries may be sufficient, even sweeping. Another justification for federal funding, say advocates, is that institutional, legal, cultural and other barriers may inhibit or prevent all parties from working together to share the costs and risks of R&D. The National Cooperative Research Act of 1984 (P.L. 98-462) and the National Cooperative Production Amendments of 1993 (P.L. 103-42) sought to spur collaborative research and manufacturing efforts by lowering legal barriers.

Opposition to expanded federal R&D efforts stem from a variety of perspectives, including those who believe that such efforts may be ineffective or counterproductive, a view held by many economists.

> The best way to deal with the many changes in demand that occur in a dynamic economy is to allow investors and workers to respond to such changes....
> Government allocation of investment that ignores market signals usually stunts growth by diverting labor and capital from more productive uses....
> An industrial policy that increases government planning, government subsidies, and international protectionism would only be a burden on our economic life and a threat to our long-term economic prosperity.[55]

---

[53] The "valley of death" is a term applied to the period in the innovation process generally between development of a laboratory prototype and its wide-scale commercial adoption. The term is an analogy intended to highlight the difficulties in overcoming barriers to innovation by evoking a comparison to the crossing of a barren desert strewn, as one writer says, with the "carcasses of great innovations."
[54] Economic Report of the President, Council of Economic Advisors, The White House, 1989. p. 223.
[55] Economic Report of the President, Council of Economic Advisors, The White House, 1984. p. 88.

Some economists assert that public R&D funding displaces private R&D investment. In his book, *The Economic Laws of Scientific Research*, Cambridge University scientist Terence Kealey argues that public R&D funding actually decreases overall R&D funding as companies reduce their R&D investments and rely on public investments.[56] Other research suggests that evidence of displacement is ambiguous.[57]

Opponents also argue that industry, not government, is best suited to make commercial technology decisions, citing the failure of some high-profile commercially-directed government efforts, such as the Concorde supersonic transport aircraft, a failed effort of the governments of the United Kingdom and France. Opponents further contend that governments — responding to political interests, not market signals — have often continued to invest in technologies — such as those supported by the U.S. synfuels program — that have been proven by markets and technological developments to be economically unsound.[58]

Libertarian opposition to increased federal R&D, such as that put forth by the Cato Institute, is grounded in a philosophy of limited government and reliance on the free market. Libertarians generally assert that markets, free from government interventions, are the most effective mechanism for allocating resources to the most promising opportunities. In their view, government interventions represent an industrial policy in which the preferences of politicians and bureaucrats are substituted for market forces and/or objective criteria. When the federal government provides direct and/or indirect financial support to a particular technology,[59] assert these advocates, it may not only provide a direct benefit to the technology —especially with respect to existing technology or alternatives — but it may also signal technology developers and investors that the technology may receive future preferential treatment by government as well. This may, as a result, skew corporate development activities and private investments toward less-promising directions producing more costly and/or less beneficial results.

Many libertarians also see government financial support for technology development as an inappropriate involuntary transfer of wealth from taxpayers to private interests — including large, highly profitable companies. The Cato Institute has labeled such efforts "corporate welfare" and has expressed concerns

---

[56] Kealey, Terrence. The Economic Laws of Scientific Research. New York: St. Martin's Press, 1996. pp. 246-247
[57] David, Paul A., Brownyn H. Hall, and Andrew A. Toole. Is Public R&D a Complement or Substitute for Private R&D? A Review of the Econometric Evidence. August 1999. [http://129.3.20.41/eps/dev/papers/ 9912/9912002.pdf]
[58] Economic Report of the President, Council of Economic Advisors, The White House, 1990. p. 117.
[59] "Technology" is used in this instance, but these arguments apply to companies and industries as well.

that such efforts may "create an unhealthy relationship between government and industry that might corrupt both."[60]

Options for federal efforts to support technology development include both direct support and indirect support:

Direct Support. There are a variety of mechanisms that the federal government might use to support downstream research. Some favor a direct approach with the federal government providing grants or loans to companies, universities, and or consortia to support research and development activities that move their work closer to commercial production. Examples of this approach include the National Institute of Standards and Technology's (NIST) Technology Innovation Program (TIP); its defunct predecessor, the Advanced Technology Program (ATP), also administered by NIST; and the multi-agency Small Business Innovation Research (SBIR) program. According to NIST, the mission of TIP is to "accelerate innovation in the United States through high-risk, high-reward research in areas of critical national need."[61] As originally conceived, ATP was intended to support the development of emerging and enabling technologies that offered the potential for significant economic and/or societal returns to the nation. The SBIR program, operating at each of the major R&D funding agencies, provides funding to help advance technology development with a goal of commercialization. (For additional information, see CRS Report 96-402, *Small Business Innovation Research Program*, and CRS Report RS22815, *The Technology Innovation Program*, both by Wendy H. Schacht.)

Indirect Support. In addition to direct funding mechanisms, a variety of indirect approaches might be used by the federal government if it chose to support additional nanotechnology research and development. The tax code could be used to increase private investment in nanotechnology companies, or to create incentives for companies to expand and accelerate their research, development, and production activities. Tax options might include general provisions to induce greater corporate investment, such as the current research and experimentation (R&E) tax credit;[62] targeted tax provisions that support a particular technology, application, industry, or sector; consumer tax deductions or credits designed to induce the purchase of targeted technologies and products, such as tax credits

---

[60] Cato Handbook on Policy, 6th Edition, The Cato Institute, 2005.
[61] NIST website. [http://www.nist.gov/public_affairs/tip.htm]
[62] The "R&E tax credit" is often referred to as the "R&D tax credit," though its provisions do not extend to development activities.

currently provided for the purchase of hybrid and flex-fuel vehicles; or incentives for the formation of capital pools to support R&D, such as favored tax treatment for research and development limited partnerships (RDLPs). (For additional information, see CRS Report RL31181, *Research and Experimentation Tax Credit: Current Status and Selected Issues for Congress*, by Gary Guenther.)

Infrastructure Development

Another option for federal support, proposed by some in industry, is increased investments in infrastructure and supporting technologies to reduce the cost of, and to accelerate, applied research and development. Candidate activities for such support include modeling, prototyping, testing, and materials characterization facilities; measurement tools and sensors; standards; reference materials; and nomenclature development.

In addition, state and local governments in the United States, as well as foreign governments, have established science, technology, and innovation parks, both specialized and general, to foster innovation. Some nanotechnology advocates believe the federal government should provide funding for the planning and development of nanotechnology-focused parks that offer land, facilities, equipment, and services to new, emerging, and established companies, and that bring together a variety of stakeholders with unique capabilities and interests.

Some in the private sector have also sought increased federal efforts to protect the intellectual property rights of inventors and companies, including increasing the speed and quality of the patent process and protecting the rights of U.S. patent holders against infringement and abuse by actors in other nations. The USPTO, an NNI-participating agency, has undertaken efforts to educate its patent examiners on nanotechnology, established a separate class for nanotechnology (Class 977), and created over 250 cross-reference sub-classes to improve the ability to search and examine nanotechnology-related patent documents.[63]

Addressing Regulatory Concerns

Environmental, health, and safety (EHS) concerns also present potential barriers to nanotechnology commercialization and U.S. competitiveness in nanotechnology. The properties of nanoscale materials — e.g., small size; high surface area-to-volume ratio; unique chemical, electric, optical, and biological

---

[63] USPTO website. [http://www1.uspto.gov/web/patents/biochempharm/crossref.htm]

characteristics — that have given rise to great hopes for beneficial applications have also given rise to concerns about their potential implications for health, safety, and the environment. EHS issues have become a specific concern of the National Nanotechnology Initiative. In FY2008, the NNI will spend $58.6 million on EHS research, accounting for about 3.9% of NNI funding. President Bush has requested $76.4 million for NNI EHS research in FY2009, or 5.0% of NNI funding. Some believe that these funding levels are too low and should amount to 10% or more of NNI funding.

The potential for adverse effects on health, safety, and the environment may discourage investment in, and development of, nanotechnology resulting from the possibility of regulations that bar products from the market or impose excessive regulatory compliance costs, and the potential for costly product liability claims and clean-up costs. If U.S. regulations are restrictive and expensive, companies may move nanotechnology research, development, and production to nations that do not impose or enforce regulations, or take a less stringent approach to regulation. Many advocates in industry, academia, and environmental non-governmental organizations believe the federal government should increase its EHS R&D investments to reduce uncertainty, inform the development of regulations, and protect the public. Regulation of nanotechnology products may fall under the authorities of several federal agencies, including the Environmental Protection Agency, Food and Drug Administration, Occupational Safety and Health Administration, and Consumer Product Safety Commission. (For additional information, see CRS Report RL34332,

Engineered Nanoscale Materials and Derivative Products: Regulatory Challenges, and CRS Report RL34118, *The Toxic Substances Control Act (TSCA): Implementation and New Challenges*, both by Linda-Jo Schierow.)

Beyond support for research and development, the federal role in a variety of other policy and programmatic activities might be strengthened. For example, some argue for the use of specialized extension centers, both university-based and independent centers, to provide technical and EHS best practices information to small and medium-size manufacturers that lack the in-house expertise and resources of larger enterprises. USDA's Agricultural Extension Service and NIST's Manufacturing Extension Partnership (MEP) may serve as possible models for such efforts.

In addition, some experts advocate efforts to create regulatory processes that can keep pace with rapid technological change and help create a more predictable environment for those investing in nanotechnology development and commercialization. Another potential regulatory barrier to nanotechnology development and commercialization is over-regulation of the export of nanotechnology and nanotechnology-related products due to their potential military applications. Such restrictions, or even the anticipation of them, might impede investment in, and development of, nanotechnology since global revenues may account for a significant share of expected return on investment. In this regard, the Department of Commerce asked the President's Export Council (PEC), a presidential advisory committee, to undertake efforts to ensure that nanotechnology products were not unnecessarily restricted from sale to other nations under export control regulations. In December 2005, the PEC sent a letter to President Bush recommending principles for the federal government's approach to export controls to maximize U.S. companies' access to global markets consistent with the protection of national and homeland security.[64] In February 2008, the Commerce Department's Bureau of Industry and Security announced its intent to establish an Emerging Technologies and Research Advisory Committee (ETRAC) comprised of representatives of research universities, government laboratories, and industry to make recommendations regarding emerging technologies, including nanotechnology.[65]

Workforce Development

Ensuring the United States has a cadre of world-class scientists, engineers, and technicians — an asset deemed critical to U.S. innovation and competitiveness —has been an enduring concern of Congress, generally,[66] and now specifically with respect to nanotechnology. Advocates for this position assert the need for federal support for curricula development, as well as scholarships and expanded efforts to encourage students to pursue associate, bachelor's and advanced degrees in nanotechnology-related disciplines.

International Coordination and Cooperation

Some nanotechnology advocates want the federal government to work with other nations to ensure a "level playing field" for nanotechnology development and commercialization (i.e., to ensure they are not put at a disadvantage by

---

[64] Transcript of President's Export Council meeting, Dirksen Senate Office Building, Washington, DC, December 6, 2005. [http://www.ita.doc.gov/td/pec/12605transcript.html].
[65] Private communication between CRS and the Bureau of Industry and Security, U.S. Department of Commerce, May 9, 2008.
[66] For additional information, see CRS Report RL34328, America COMPETES Act: Programs, Funding, and Selected Issues, by Deborah D. Stine.

government subsidization of their foreign competitors, less stringent regulatory standards, fewer worker protections, and/or imposition of tariffs and non-tariff trade barriers), to develop common international standards and nomenclature, to harmonize regulations, and to open markets for nanotechnology products.

Reassessing and Realigning Resources

As discussed above, the federal government is engaged in fostering the advancement of nanotechnology across a broad range of activities, including: conducting and supporting nanotechnology R&D; seeking to address environmental, health, and safety issues; preparing students and workers for nanotechnology job opportunities through investments in education and training; fostering public understanding and engagement; coordinating and cooperating with other nations; and promoting the development of standards, nomenclature, and reference materials. These activities involve substantial investments of capital, personnel, facilities, equipment, and other resources.

Many NNI activities have developed over time to address new challenges and opportunities as the NNI advanced. Resource allocation decisions have been made piecemeal, generally without consideration for alternative uses of the resources. Over time, such an approach may produce a portfolio of activities that is out of balance with current needs. More than seven years into the NNI, some observers believe that reassessing and realigning resources with opportunities and challenges would improve the efficiency and effectiveness of federal investments and activities.

However, there are substantial barriers to such an effort. First, the NNI is not funded centrally, but rather is an aggregation of the resources provided to agencies to meet their mission requirements. Moving funds from one program or agency to another might meet with resistance within agencies, between agencies, or from the Congressional appropriations subcommittees with jurisdiction for these programs and agencies. Second, agencies participate in the NNI on a voluntary basis. If it appears that participation in the NNI might reduce funding, an agency may choose to no longer participate or may not classify its activities as nanotechnology. Third, the NNI seeks to meet multiple goals, including scientific leadership, meeting agency mission requirements and national needs, and fostering U.S. commercial leadership. No relative values have been explicitly set for these and other goals making comparative resource allocation choices subjective.

The National Research Council (NRC)[67] and the National Nanotechnology Advisory Panel (NNAP)[68] have each conducted assessments of the NNI at the direction of Congress as specified in the 21st Century Nanotechnology Research and Development Act. The act requires assessments to be performed triennially by the NRC and biennially by the NNAP. Past assessments have addressed U.S. competitiveness in nanotechnology as part of wider reviews. To clarify the U.S. competitive position, Congress could opt to direct the NRC, NNAP, or the U.S. Government Accountability Office to conduct a focused assessment of: the effectiveness of the NNI in achieving the global technological leadership, commercialization, and national competitiveness goals established under the act; whether the current portfolio of NNI resources and activities are appropriately balanced; and whether additional resources activities may be required to achieve these objectives.

CONCLUDING OBSERVATIONS

Nanotechnology is expected by many to deliver significant economic and societal benefits. The United States launched the first national nanotechnology initiative in 2000, but has since been joined by more than 60 other nations. Tens of billions of dollars have been invested in nanotechnology research and development over the past eight years by governments, companies, and investors.

While it has been estimated that there are more than 600 nanotechnology products on the market today, most involve incremental improvements to existing products. Much of the investment has been focused on fundamental research to gain scientific understanding of nanoscale phenomena and processes, and to learn how to manipulate matter at the nanoscale. These investments are expected by many to deliver revolutionary changes in products and industries with implications for global technological, economic, and military leadership. The potential implications of nanotechnology, coupled with the substantial sustained investments, have raised concerns and interest in the U.S. competitive position in nanotechnology.

The data typically used to assess technological competitiveness in mature industries — e.g., revenues, market share, trade — is not available to assess the

---

[67] The National Research Council functions under the auspices of the National Academy of Sciences, the National Academy of Engineering, and the Institute of Medicine.
[68] In July 2004, President Bush designated the President's Council of Advisors on Science and Technology to serve as the NNAP by issuing Executive Order 13349, Amending Executive Order 13226 To Designate the President's Council ofAdvisors on Science and Technology To Serve as the National Nanotechnology Advisory Panel.

U.S. position in nanotechnology because it is a new technology, commercial products are just beginning to enter the market in a significant manner, and it is incorporated in wide array of products across many industries. Accordingly, the federal government currently does not collect this data on nanotechnology, nor do other nations. The number of nanotechnology products in the marketplace is increasing quickly though. Congress may elect to ask federal agencies to assess what data (e.g. economic, labor force, students) would be useful in formulating federal policies and making resource allocation decisions and direct federal statistical agencies to collect, analyze, and make public such data. The federal government may also seek to foster data collection efforts in other nations.

In the absence of such data, assessments of nanotechnology depend largely on alternative indicators, such as inputs (e.g., public and private investments) and noneconomic outputs (e.g., scientific papers, patents). By these measures, the United States appears to lead all other nations in nanotechnology, though the U.S. lead in this field may not be as large as it has been in previous emerging technology areas. This is due to increased investments and capabilities of many nations based on recognition that technological leadership and commercialization are primary paths to increased economic growth, improved standards of living, and job creation.

Nevertheless, these alternative indicators may not present an accurate view of technological leadership and economic competitiveness for many reasons. Nor does national technological leadership alone guarantee that the economic value produced by nanotechnology innovations will be captured within a nation's borders. In today's global economy, companies have the option of locating work — e.g. research, development, design, engineering, manufacturing, product support — where it can be done most effectively.

A variety of federal policy issues may affect the development and commercialization of nanotechnology in the United States, including the magnitude and focus of research and development efforts, the regulatory environment, and science and engineering workforce development. Some support an active federal approach; others believe that a more limited federal involvement is likely to be more successful and equitable. In addition to these factors, U.S. competitiveness in nanotechnology will depend not just on the efforts of the United States, but also on the speed and efficacy of foreign nanotechnology development efforts.

Congress established a legislative foundation for some of the activities of the National Nanotechnology Initiative and to address key issues associated with nanotechnology thorough enactment of the 21st Century Nanotechnology Research and Development Act, 2003. The act provided funding authorizations for five NNI agencies through FY2008. Action is being considered in both the House and Senate on possible amendments to and reauthorization of the program. Congress may opt to address some or many of the issues identified in this paper in the course of deliberation on the reauthorization of this act or, alternatively, in separate legislation.

Chapter 2

NANOTECHNOLOGY: A POLICY PRIMER[*]

John F. Sargent
Science and Technology Policy Resources,
Science, and Industry Division

SUMMARY

Nanoscale science, engineering and technology — commonly referred to collectively as nanotechnology — is believed by many to offer extraordinary economic and societal benefits. Congress has demonstrated continuing support for nanotechnology and has directed its attention primarily to three topics that may affect the realization of this hoped for potential: federal research and development (R&D) in nanotechnology; U.S. competitiveness; and environmental, health, and safety (EHS) concerns. This report provides an overview of these topics — which are discussed in more detail in current and upcoming CRS reports — and two others: nanomanufacturing and public understanding of and attitudes toward nanotechnology.

The development of this emerging field has been fostered by significant and sustained public investments in nanotechnology R&D. Nanotechnology R&D is directed toward the understanding and control of matter at dimensions of roughly 1 to 100 nanometers. At this size, the properties of matter can differ in fundamental and potentially useful ways from the properties of individual atoms and molecules and of bulk matter. Since the launch of the National Nanotechnology Initiative (NNI) in 2000, Congress has appropriated approximately $8.4 billion for nanotechnology R&D. More than 60 nations have established similar programs. In 2006 alone, total global public R&D investments reached an estimated $6.4 billion, complemented by an estimated private sector

---
[*] This report excerpted from CRS report #106154, May 20, 2008

investment of $6.0 billion. Data on economic outputs that are used to assess competitiveness in mature technologies and industries, such as revenues and market share, are not available for assessing nanotechnology. Alternatively, data on inputs (e.g., R&D expenditures) and non-financial outputs (e.g. scientific papers, patents) may provide insight into the current U.S. position and serve as bellwethers of future competitiveness. By these criteria, the United States appears to be the overall global leader in nanotechnology, though some believe the U.S. lead may not be as large as it has been for previous emerging technologies.

Some research has raised concerns about the safety of nanoscale materials. There is general agreement that more information on EHS implications is needed to protect the public and the environment; to assess and manage risks; and to create a regulatory environment that fosters prudent investment in nanotechnology-related innovation. Nanomanufacturing — the bridge between nanoscience and nanotechnology products — may require the development of new technologies, tools, instruments, measurement science, and standards to enable safe, effective, and affordable commercial-scale production of nanotechnology products. Public understanding and attitudes may also affect the environment for R&D, regulation, and market acceptance of products incorporating nanotechnology.

In 2003, Congress enacted the 21st Century Nanotechnology Research and Development Act providing a legislative foundation for some of the activities of the NNI, addressing concerns, establishing programs, assigning agency responsibilities, and setting authorization levels. Both the House of Representatives and the Senate remain actively engaged in the NNI, holding hearings in 2007 and 2008 related to possible amendments to, and reauthorization of, the act. Policy issues related to the NNI may be addressed in this process or through separate legislation.

OVERVIEW

Congress continues to demonstrate interest in and support for nanotechnology due to what many believe is its extraordinary potential for delivering economic growth, high-wage jobs, and other societal benefits to the nation. To date, the Science Committee in the House and Senate Committee on Commerce have directed their attention primarily to three topics that may affect the United States' realization of this hoped for potential: federal research and development (R&D) investments under the National Nanotechnology Initiative (NNI); U.S. international competitiveness; and environmental, health, and safety (EHS) concerns. This report provides a brief overview of these topics — which are discussed in greater detail in current and upcoming CRS reports[1] — and two other

---

[1] For additional information on these issues, see CRS Report RL34401, The National Nanotechnology Initiative: Overview, Reauthorization, and Appropriations Issues, and CRS Report RL34493, Nanotechnology and U.S. Competitiveness, both by John F. Sargent, and CRS

subjects of interest to Congress: nanomanufacturing and public attitudes toward, and understanding of, nanotechnology.

Nanotechnology research and development is directed toward the understanding and control of matter at dimensions of roughly 1 to 100 nanometers. At this size, the physical, chemical, and biological properties of materials can differ in fundamental and potentially useful ways from the properties of individual atoms and molecules, on the one hand, or bulk matter, on the other hand.

In 2000, President Clinton launched the NNI to coordinate federal R&D efforts and promote U.S. competitiveness in nanotechnology. Congress first funded the NNI in FY2001 and has provided increased appropriations for nanotechnology R&D in each subsequent year. In 2003, Congress enacted the 21st Century Nanotechnology Research and Development Act (P.L. 108-153). The act provided a statutory foundation for the NNI, established programs, assigned agency responsibilities, authorized funding levels, and initiated research to address key issues.

Federal R&D investments are focused on advancing understanding of fundamental nanoscale phenomena and on developing nanomaterials, nanoscale devices and systems, instrumentation, standards, measurement science, and the tools and processes needed for nanomanufacturing. NNI appropriations also fund the construction and operation of major research facilities and the acquisition of instrumentation. Finally, the NNI supports research directed at identifying and managing potential environmental, health, and safety impacts of nanotechnology, as well as its ethical, legal and societal implications.

Most current applications of nanotechnology are evolutionary in nature, offering incremental improvements in existing products and generally modest economic and societal benefits. For example, nanotechnology is being used: in automobile bumpers, cargo beds, and step-assists to reduce weight, increase resistance to dents and scratches, and eliminate rust; in clothes to increase stain- and wrinkle-resistance; and in sporting goods, such as baseball bats and golf clubs, to improve performance.

In the longer term, nanotechnology may deliver revolutionary advances with profound economic and societal implications. Potential applications discussed by the technology's proponents involve various degrees of speculation and varying time-frames. The examples below suggest areas where such possible

---

Report RL34332, Engineered Nanoscale Materials and Derivative Products: Regulatory Challenges, by Linda-Jo Schierow. An upcoming CRS report will address nanotechnology environmental, health, and safety issues.

revolutionary advances may emerge, and early research and development efforts that may provide insights into how such advances may be achieved.

- **Detection and treatment technologies for cancer and other deadly diseases.** Current nanotechnology disease detection efforts include the development of sensors that can identify biomarkers, such as altered genes, that may provide an early indicator of cancer. One approach uses carbon nanotubes and nanowires to identify the unique molecular signals of cancer biomarkers. Another approach uses nanoscale cantilevers — resembling a row of diving boards — treated with molecules that bind only with cancer biomarkers. When these molecules bind, the additional weight bends the cantilevers indicating the presence and concentration of these biomarkers. Nanotechnology holds promise for showing the presence, location, and/or contours of cancer, cardiovascular disease, or neurological disease. Current R&D efforts employ metallic, magnetic, and polymeric nanoparticles with strong imaging characteristics attached to an antibody or other agent that binds selectively with targeted cells. The imaging results can be used to guide surgical procedures and to monitor the effectiveness of non-surgical therapies in killing the disease or slowing its growth. Nanotechnology may also offer new cancer treatment approaches. For example, nanoshells with a core of silica and an outer metallic shell can be engineered to concentrate at cancer lesion sites. Once at the sites, a harmless energy source (such as near-infrared light) can be used to cause the nanoshells to heat, killing the cancer cells they are attached to. Another treatment approach targets delivery of tiny amounts of a chemotherapy drug to cancer cells. In this approach the drug is encapsulated inside a nanoshell that is engineered to bind with an antigen on the cancer cell. Once bound, the nanoshell dissolves, releasing the chemotherapy drug killing the cancer cell. Such a targeted delivery approach could reduce the amount of chemotherapy drug needed to kill the cancer cells, reducing the side effects of chemotherapy.[2]
- **Clean, inexpensive, renewable power through energy creation, storage, and transmission technologies.** Nanoscale semiconductor catalysts and additives show promise for improving the production of hydrogen from water using sunlight. The optical properties of these nanoscale catalysts allow the process to use a wider spectrum of sunlight. Similarly, nanostructured photovoltaic devices (e.g. solar panels) may

---

[2] National Cancer Institute website. [http://nano.cancer.gov/resource_center/tech_backgrounder.asp]

improve the efficiency of converting sunlight into electricity by using a wider spectrum of sunlight. Improved hydrogen storage, a key challenge in fuel cell applications, may be achieved by tapping the chemical properties and large surface area of certain nanostructured materials. In addition, carbon nanotube fibers have the potential for reducing energy transmission losses from approximately 7% (using copper wires) to 6% (using carbon nanotube fibers), an equivalent annual energy savings in the United States of 24 million barrels of oil.[3]

- **Universal access to clean water.** Nanotechnology water desalination and filtration systems may offer affordable, scalable, and portable water filtration systems. Filters employing nanoscale pores work by allowing water molecules to pass through, but prevent larger molecules, such as salt ions and other impurities (e.g. bacteria, viruses, heavy metals, and organic material), from doing so. Some nanoscale filtration systems also employ a matrix of polymers and nanoparticles that serve to attract water molecules to the filter and to repel contaminants.[4]

- **High-density memory devices.** A variety of nanotechnology applications may hold the potential for improving the density of memory storage. For example, IBM has demonstrated the potential to create high-density memory devices (with an estimated storage capacity of 1 terabyte per square inch) by storing information mechanically using thermal-mechanical nanoscale probes to punch nanoscale indentations into a thin plastic film. The probes can be used to read and write data in parallel.[5]

- **Higher crop yield and improved nutrition**. Higher crop yield might be achieved using nanoscale sensors that detect the presence of a virus or disease-infecting particle. Early, location-specific detection may allow for rapid and targeted treatment of affected areas, increasing yield by preventing losses.[6] Nanotechnology also offers the potential for improved nutrition. Some companies are exploring the development of

---

[3] Nanoscience Research for Energy Needs, Nanoscale Science, Engineering, and Technology Subcommittee, National Science and Technology Council, The White House, December 2004.
[4] Abraham, M. "Today's Seawater is Tomorrow's Drinking Water," University of California at Los Angeles, November 6, 2006
[5] Vettiger, P. "The 'millipede' - nanotechnology entering data storage," IEEE Transactions on Nanotechnology, March 2002. Vol 1. Issue 1. pp 29-55.
[6] Nanoscale Science and Engineering for Agriculture and Food Systems, draft report on the National Planning Workshop, submitted to the Cooperative State Research, Education, and Extension Service of the U.S. Department of Agriculture, July 2003.

nanocapsules that release nutrients targeted at specific parts of the body at specific times.[7]

- **Self-healing materials.** Nanotechnology may offer approaches that enable materials to "self-heal" by incorporating, for example, nanocontainers of a repair substance (e.g., an epoxy) throughout the material. When a crack or corrosion reaches a nanocontainer, the nanocontainer could be designed to open and release its repair material to fill the gap and seal the crack.[8]
- **Sensors that can warn of minute levels of toxins and pathogens in air, soil or water.** Microfluidic and nanocantilever sensors (discussed earlier) may be engineered to detect specific pathogens (e.g. bacteria, virus) or toxins (e.g., sarin gas, hydrogen cyanide) by detecting their unique molecular signals or through selective binding with an engineered nanoparticle.
- *Environmental remediation of contaminated sites.* The high surface-to-volume ratio, high reactivity, and small size of some nanoscale particles (e.g. nanoscale iron) may offer more effective and less costly solutions to environmental contamination. By injecting engineered nanoparticles into the ground, these characteristics can be employed to enable the particles to move more easily through a contaminated site and bond more readily with targeted contaminants.[9]

Nanotechnology is also expected to make substantial contributions to federal missions such as national defense, homeland security, and space exploration and commercialization.

U.S. private sector nanotechnology R&D is now estimated to be twice that of public funding. In general, the private sector's efforts are focused on translating fundamental knowledge and prototypes into commercial products; developing new applications incorporating nanoscale materials; and developing technologies, methods, and systems for commercial-scale manufacturing.

Many other nations and firms around the world are also making substantial investments in nanotechnology to reap its potential benefits. With so much potentially at stake, some members of Congress have expressed interest and concerns about the U.S. competitive position in nanotechnology R&D and success in translating R&D results to commercial products. This has led to an increased

---

[7] Wolfe, Josh. "Safer and Guilt-Free Nano Foods," Forbes.com, August 10, 2005.
[8] Berger, Michael. "Nanomaterial heal thyself," Nanowerk Spotlight, June 13, 2007.
[9] EPA website. [http://es.epa.gov/ncer/nano/research/nano_remediation.html]

focus on potential barriers to commercialization efforts, including the readiness of technologies, systems, and processes for large-scale nanotechnology manufacturing; potential environmental, health, and safety (EHS) effects of nanoscale materials; public understanding and attitudes toward nanotechnology; and other related issues. Both the House of Representatives and the Senate have held hearings in 2008 on amending the 21st Century Nanotechnology Research and Development Act. This report provides a macro-level view of federal R&D investments, U.S. competitiveness in nanotechnology, and EHS-related issues.

THE NATIONAL NANOTECHNOLOGY INITIATIVE

President Clinton launched the National Nanotechnology Initiative in 2000, establishing a multi-agency program to coordinate and expand federal efforts to advance the state of nanoscale science, engineering, and technology, and to position the United States to lead the world in its development and commercialization. The NNI is comprised of 13 federal agencies that receive appropriations to conduct and fund nanotechnology R&D and 12 other federal agencies with responsibilities for health, safety, and environmental regulation; trade; education; training; intellectual property; international relations; and other areas that might affect nanotechnology. EPA both conducts R&D and has regulatory responsibilities.

Congress has played a central role in the NNI, providing appropriations for the conduct of nanotechnology R&D (discussed below), establishing programs, and creating a legislative foundation for some of the activities of the NNI through enactment of the 21st Century Nanotechnology Research and Development Act of 2003. The act also authorized appropriations FY2005 through FY2008 for five NNI agencies — the National Science Foundation (NSF), Department of Energy (DOE), National Aeronautics and Space Administration (NASA), Department of Commerce (DOC) National Institute of Standards and Technology (NIST), and Environmental Protection Agency (EPA).

Structure

The NNI is coordinated within the White House through the National Science and Technology Council (NSTC) Nanoscale Science, Engineering, and Technology (NSET) subcommittee. The NSET subcommittee is comprised of representatives from 25 federal agencies, White House Office of Science and

Technology Policy (OSTP) and Office of Management and Budget.[10] The NSETsubcommittee has established several working groups, including the National Environmental and Health Implications (NEHI), National Innovation and Liaison with Industry (NILI), Global Issues in Nanotechnology (GIN), Nanomanufacturing, and Nanotechnology Public Engagement and Communications (NPEC) working groups. The National Nanotechnology Coordination Office (NNCO) provides administrative and technical support to the NSET subcommittee.

Funding

Funding for the NNI is provided through appropriations to each of the NNI-participating agencies. The NNI has no centralized funding. Overall NNI funding is calculated by aggregating the nanotechnology-related expenditures of each NNI agency. Funding remains concentrated in the original six NNI agencies[11] which account for 98.3% of NNI funding in FY2008. The NNI funds fundamental and applied nanotechnology R&D, multidisciplinary centers of excellence, and key research infrastructure. It also supports efforts to address societal implications of nanotechnology, including ethical, legal, EHS, and workforce issues.

For FY2008, Congress appropriated an estimated $1.491 billion for nanotechnology R&D, more than triple the $464 million federal investment in 2001. In total, Congress appropriated approximately $8.4 billion for the NNI since FY2001. President Bush has requested $1.527 billion for nanotechnology R&D in FY2009, a 2.3% increase above the estimated FY2008 funding level. The chronology of NNI funding is detailed in Table 1.

---

[10] NSET subcommittee members include Bureau of Industry and Security, DOC; Consumer Product Safety Commission; Cooperative State Research, Education, and Extension Service, Department of Agriculture (USDA); Department of Defense (DOD); Department of Education; DOE; Department of Homeland Security; Department of Justice; Department of Labor; Department of State; Department of Transportation; Department of the Treasury; EPA; Food and Drug Administration; Forest Service, USDA; Intelligence Technology Innovation Center; International Trade Commission; NASA; National Institutes of Health (NIH), Department of Health and Human Services (DHHS); National Institute for Occupational Safety and Health, Centers for Disease Control and Prevention, (DHHS); NIST, DOC; NSF; Nuclear Regulatory Commission; U.S. Geological Survey, Department of the Interior; and U.S. Patent and Trademark Office, DOC.

[11] The original six agencies were the NSF, DOD, DOE, NIST, NASA, and NIH.

Table 1. NNI Funding, by Agency
(in millions of current dollars)

Agency	FY 2001 Actual	FY 2002 Actual	FY 2003 Actual	FY 2004 Actual	FY 2005 Actual	FY 2006 Actual	FY 2007 Actual	FY 2008 Estimate	FY 2009 Request
DOD	125	224	220	291	352	424	450	487	431
NSF	150	204	221	256	335	360	389	389	397
DOE	88	89	134	202	208	231	236	251	311
NIH (DHHS)	40	59	78	106	165	192	215	226	226
NIST (DOC)	33	77	64	77	79	78	88	89	110
NASA	22	35	36	47	45	50	20	18	19
EPA	5	6	5	5	7	5	8	10	15
Other Agencies	1	3	2	5	9	13	19	21	18
TOTAL[b]	464	697	760	989	1,200	1,351	1,425	1,491	1,527

Source: NNI website. [http://www.nano.gov/html/about/funding.html]

[a.] According to NSTC, the Department of Defense budgets for FY2006, FY2007, and FY2008 include Congressionally directed funding outside the NNI plan. The extent to which such funding is included or not included in reporting of funding in earlier fiscal years is uncertain.

[b.] Numbers may not add due to rounding of agency budget figures.

SELECTED ISSUES

U.S. Competitiveness

Nanotechnology is largely still in its infancy. Accordingly, measures such as revenues, market share, and global trade statistics — which are often used to assess and track U.S. competitiveness in other more mature technologies and industries — are not available for assessing the relative U.S. position internationally in nanotechnology. To date, the federal government does not collect data on nanotechnology-related revenues, trade or employment, nor is comparable international government data available. Nevertheless, many experts believe that the United States is the global leader in nanotechnology. However, some of these experts believe that in contrast to many previous emerging technologies — such as semiconductors, satellites, software, and biotechnology — the U.S. lead is narrower, and the investment level, scientific and industrial infrastructure, technical capabilities, and science and engineering workforces of other nations are more substantial than in the past.

In the absence of comprehensive and reliable economic output data (e.g., revenues, market share, trade), indicators such as inputs (e.g., public and private research investments) and non-financial outputs (e.g., scientific papers, patents) have been used to gauge a nation's competitive position in emerging technologies. By these measures (discussed below), the United States appears to lead the world, generally, in nanotechnology. However, R&D investments, scientific papers, and patents may not provide reliable indicators of the United States' current or future competitive position. Scientific and technological leadership may not necessarily result in commercial leadership or national competitiveness for a variety of reasons:

- Basic research in nanotechnology may not translate into viable commercial applications.
- Basic research is generally available to all competitors.
- U.S.-based companies may conduct production and other work outside of the United States.
- U.S.-educated foreign students may return home to conduct research and create new businesses.
- U.S. companies with leading-edge nanotechnology capabilities and/or intellectual property may be acquired by foreign competitors.
- U.S. policies or other factors may prohibit nanotechnology commercialization, make it unaffordable, or make it less attractive than foreign alternatives.
- Aggregate national data may be misleading as countries may establish global leadership in niche areas of nantoechnology.

With these caveats, the following section reviews input and non-economic output measures as indicators of the U.S. competitive position in nanotechnology.

Global Funding

The United States has led, and continues to lead, all nations in known public investments in nanotechnology R&D, though the estimated U.S. share of global public investments has fallen as other nations have established similar programs and increased funding. Using a currency exchange rate comparison, the United States ranks ahead of all other nations, with federal and state investments of $1.78 billion in 2006 (27.8% of global public R&D investments), followed by Japan ($975 million, 15.2%) and Germany ($563 million, 8.8%). When national investments are adjusted using purchasing power parity (PPP) exchange rates,[12]

---

[12] Purchasing power parity exchange rates seek to equalize the purchasing power of currencies in different countries for a given basket of goods and/or services.

the United States remains the world leader, but China ranked second in public nanotechnology spending in 2005.[13]

Private investments in nanotechnology R&D come from two primary sources, corporations and venture capital investors. On a PPP comparison basis, the United States led the world in 2006 in corporate R&D investments in nanotechnology with an estimated $1.9 billion investment, followed by Japan with $1.7 billion. In total, U.S. and Japan-based companies accounted for nearly three-fourths of global corporate investment in nanotechnology R&D in 2006. China ranks fifth in corporate investment, accounting for approximately 3% of global private R&D investments.[14] Lux Research, an emerging technologies consulting firm, estimates that global nanotechnology venture capital investment in 2007 was $702 million, of which $632 million went to U.S.-based firms.[15]

Scientific Papers

The quantity of peer-reviewed scientific papers is considered by some to be an indicator of a nation's scientific leadership. A study by the National Bureau of Economic Research in 2005 reported that the U.S. share was a world-leading 24%, but that this represented a decline from approximately 40% in the early 1990s, concluding:

> Taken as a whole these data confirm that the strength and depth of the American science base points to the United States being the dominant player in nanotechnology for some time to come, while the United States also faces significant and increasing international competition.[16]

Patents

Patent counts — assessments of how many patents are issued to individuals or institutions of a particular country — are frequently used to assess technological competitiveness. By this measure, the U.S. competitive position appears to be strong. A 2007 U.S. Patent and Trademark Office analysis of patents in the United States and in other nations stated that U.S.-origin inventors and assignees/owners have:

- the most nanotechnology-related U.S. patents by a wide margin;

---

[13] Profiting from International Nanotechnology, Lux Research, Inc., December 2006.
[14] Profiting from International Nanotechnology, Lux Research, Inc., December 2006.
[15] Personal communication between CRS and Lux Research, April 28, 2008.
[16] Zucker, L.G. and M.R. Darby. "Socio-Economic Impact of Nanoscale Science: Initial Results and Nanobank," National Bureau of Economic Research, March 2005.

- the most nanotechnology-related patent publications globally, but by a narrower margin (followed closely by Japan); and
- the most nanotechnology-related inventions that have patent publications in three or more countries, 31.7%, followed by Japan (26.9%), Germany (11.3%), Korea (6.6%), and France (3.6%).[17]

Environmental, Health, and Safety Implications

Key policy issues associated with U.S. competitiveness in nanotechnology include environmental, health, and safety (EHS) concerns, nanomanufacturing, and public understanding and attitudes. EHS concerns include both direct consequences for health, safety, and the environment, and how uncertainty about EHS implications and potential regulatory responses might affect U.S. competitiveness. One such effect might be the discouragement of investment in nanotechnology due to the possibility of regulations that might bar products from the market, impose high regulatory compliance costs, or result in product liability claims and clean-up costs.

Some of the unique properties of nanoscale materials — e.g., small size, high surface area-to-volume ratio — have given rise to concerns about their potential implications for health, safety, and the environment. While nanoscale particles occur naturally and as incidental by-products of other human activities (e.g., soot),[18] EHS concerns have been focused primarily on nanoscale materials that are intentionally engineered and produced.

Much of the public dialogue about risks associated with nanotechnology has focused on carbon nanotubes (CNTs) and other fullerenes (molecules formed entirely of carbon atoms in the form of a hollow sphere, ellipsoid, or tube) since they are currently being manufactured and are among the most promising nanomaterials. These concerns have been amplified by some research on the effects of CNTs onanimals, and on animal and human cells. For example, researchers have reported that carbon nanotubes inhaled by mice can cause lung tissue damage;[19] that buckyballs (spherical fullerines) caused brain damage in

---

[17] Eloshway, Charles. "Nanotechnology Related Issues at the U.S. Patent and Trademark Office," Workshop on Intellectual Property Rights in Nanotechnology: Lessons from Experiences Worldwide, Brussels, Belgium, April 2007.

[18] Some naturally occurring nanoparticles cause adverse health effects. Studies on the effects of naturally occurring particles are numerous and inform R&D on engineered nanoparticles.

[19] Lam, C.; James, J.T.; McCluskey, R.; and Hunter, R. "Pulmonary toxicity of single-wall carbon nanotubes in mice 7 and 90 days after intratracheal instillation," Toxicological Sciences, September 2003. Vol 77. No. 1. pp 126-134.

fish;[20] and that buckyballs can accumulate within cells and potentially cause DNA damage.[21] On the other hand, some research has found CNTs and fullerenes to be non-toxic. In addition, work at Rice University's Center for Biological and Environmental Nanotechnology conducted in 2005 found cell toxicity of CNTs to be low and that toxicity can be reduced further through simple chemical changes to the CNT's surface.[22]

Among the potential EHS benefits of nanotechnology are applications that may reduce energy consumption, pollution, and greenhouse gas emissions; remediate environmental damage; cure, manage, or prevent deadly diseases; and offer new materials that protect against impacts, self-repair to prevent catastrophic failure, or change in ways that provide protection and medical aid to soldiers on the battlefield.

Potential EHS risks of nanoscale particles in humans and animals depend in part on their potential to accumulate, especially in vital organs such as the lungs and brain, that might harm or kill, and diffusion in the environment that might harm ecosystems. For example, several products on the market today contain nanoscale silver, an effective antibacterial agent. Some scientists have raised concerns that the dispersion of nanoscale silver in the environment could kill microbes that are vital to ecosystems.

Like nanoscale silver, other nanoscale particles might produce both positive and negative effects. For example, some nanoscale particles have the potential to penetrate the blood-brain barrier, a structure that protects the brain from harmful substances in the blood. Currently, the barrier hinders the delivery of therapeutic agents to the brain.[23] The characteristics of some nanoscale materials may allow pharmaceuticals to be developed to purposefully and beneficially cross the blood-brain barrier and deliver medicine directly to the brain to treat, for example, a brain tumor. Alternatively, other nanoscale particles might unintentionally pass through this barrier and harm humans and animals.

There is widespread uncertainty about the potential EHS implications of nanotechnology. A survey of business leaders in the field of nanotechnology indicated that nearly two-thirds believe that "the risks to the public, the workforce, and the environment due to exposure to nano particles are 'not known,'"

---

[20] Oberdorster, Eva. "Manufactured Nanomaterials (Fullerenes, C60) Induce Oxidative Stress in the Brain of Juvenile Largemouth Bass," Environmental Health Perspectives, July 2004. Vol. 112. No. 10.
[21] "Understanding Potential Toxic Effects of Carbon-Based Nanomaterials," Nanotech News, National Cancer Institute Alliance for Nanotechnology in Cancer, July 10, 2006.
[22] "Modifications render carbon nanotubes nontoxic," press release, Rice University, October 2005.
[23] "Blood-Brain Barrier Breached by New Therapeutic Strategy," press release, National Institutes of Health, June 2007.

and 97% believe that it is very or somewhat important for the government to address potential health effects and environmental risks that may be associated with nanotechnology.[24]

Many stakeholders believe that concerns about potential detrimental effects of nanoscale materials and products on health, safety, and the environment — both real and perceived — must be addressed for a variety of reasons, including:

- protecting and improving human health, safety, and the environment;
- enabling accurate and efficient risk assessments, risk management, and cost-benefit trade-offs;
- creating a predictable, stable, and efficient regulatory environment that fosters investment in nanotechnology-related innovation;
- ensuring public confidence in the safety of nanotechnology research, engineering, manufacturing, and use;
- preventing the negative consequences of a problem in one application area of nanotechnology from harming the use of nanotechnology in other applications due to public fears, political interventions, or an overly-broad regulatory response; and
- ensuring that society can enjoy the widespread economic and societal benefits that nanotechnology may offer.

Policy issues associated with EHS impacts of nanotechnology include magnitude, timing, foci, and management of the federal investment in EHS research; adequacy of the current regulatory structures to protect public health and the environment; and cooperation with other nations engaged in nanotechnology R&D to ensure all are doing so in a responsible manner.

Nanomanufacturing

Securing the economic benefits and societal promise of nanotechnology requires the ability to translate knowledge of nanoscience into market-ready nanotechnology products. Nanomanufacturing is the bridge connecting nanoscience and nanotechnology products. Although some nanotechnology products have already entered the market, these materials and devices have tended to require only incremental changes in manufacturing processes. Generally, they

---

[24] "Survey of U.S. Nanotechnology Executives," Small Times Magazine and the Center for Economic and Civic Opinion at the University of Massachusetts-Lowell, Fall 2006.

are produced in a laboratory environment in limited quantities with a high-degree of labor intensity, high variability, and high costs. To make their way into safe, reliable, effective, and affordable commercial-scale production in a factory environment may require the development of new and unique technologies, tools, instruments, measurement science, and standards for nanomanufacturing.

Public Attitudes and Understanding

What the American people know about nanotechnology and the attitudes that they have toward it may affect the environment for research and development (especially support for public R&D funding), regulation, market acceptance of products incorporating nanotechnology, and, perhaps, the ability of nanotechnology to weather a negative event such as an accident or spill.

In 2007, the Woodrow Wilson International Center for Scholars' Project on Emerging Nanotechnologies (PEN) reported results of a nationwide poll of adults that found more than 42% had "heard nothing at all" about nanotechnology, while only 6% said they had "heard a lot." In addition, more than half of those surveyed felt they could not assess the relative value of nanotechnology's risks and benefits. Among those most likely to believe that benefits outweigh risks were those earning more than $75,000 per year, men, people who had previously heard "some" or "a lot" about nanotechnology, and those between the ages of 35 and 64. Alternatively, among those most likely to believe that the risks of nanotechnology outweigh benefits include people earning $30,000 or less; those with a high school diploma or less; women; racial and ethnic minorities; and those between the ages of 18 and 34 or over age 65.[25]

The PEN survey found a strong positive correlation between familiarity with and awareness of nanotechnology and perceptions that benefits will outweigh risks. However, the survey data also indicate that communicating with the public about nanotechnology in the absence of clear, definitive answers to EHS questions could create a higher level of uncertainty, discomfort, and opposition.

Congress expressed its belief in the importance of public engagement in the 21st Century Nanotechnology Research and Development Act of 2003 (15 U.S.C. §§7501 et seq.). The act calls for public input and outreach to be integrated into the NNI's efforts. The NNI has sought to foster public understanding through a variety of mechanisms, including written products, speaking engagements, a web-

---

[25] "Awareness of and Attitudes Toward Nanotechnology and Federal Regulatory Agencies: A Report of Findings," survey by Peter D. Hart Research Associates, Inc., for the Project on Emerging Nanotechnologies, September 2007.

based information portal (nano.gov), informal education, and efforts to establish dialogues with key stakeholders and the general public. In addition, the NSET subcommittee has established a Nanotechnology Public Engagement and Communications working group to develop approaches by which the NNI can communicate more effectively with the public.

INDEX

A

access, 6, 9, 22, 29, 39
accounting, 15, 20, 28, 45
additives, 38
administrative, 42
adults, 49
age, 49
agent, 38, 47
agents, 47
aggregation, 4, 30
agricultural, 6
aid, viii, 2, 47
air, 6, 40
alternative, 30, 32
alternatives, 11, 25, 44
amendments, ix, 33, 36
analysts, 3
animals, 47
antibacterial, 47
antibody, 38
antigen, 38
application, 5, 21, 26, 48
applied research, 9, 13, 23, 27
appropriations, 4, 5, 7, 30, 37, 41, 42
argument, 24
assessment, vii, 2, 3, 9, 31
atoms, viii, 5, 35, 37, 46
ATP, 26

attitudes, viii, ix, 11, 35, 36, 37, 41, 46, 49
auto parts, 5
availability, viii, 2, 11
awareness, 49

B

bacteria, 39, 40
barrier, 28, 47
barriers, 22, 24, 27, 30, 41
basic research, 9, 22
batteries, 15
battery, 15
Belgium, 20, 46
benefits, vii, viii, 1, 2, 5, 7, 22, 24, 31, 35, 36, 37, 40, 47, 48, 49
binding, 40
biomarkers, 38
biotechnology, 8, 43
bipartisan, 3
blood, 47
blood-brain barrier, 47
brain, 46, 47
brain damage, 46
brain tumor, 47
broad spectrum, 11
Brussels, 20, 46
Bureau of Industry and Security, 29, 42
business environment, 12, 22

by-products, 46

C

Canada, 14, 18, 20
cancer, 6, 38
cancer cells, 38
cancer treatment, 38
capacity, 39
carbon, 15, 38, 39, 46, 47
Carbon, 15
carbon atoms, 46
carbon nanotubes, 15, 38, 46, 47
cardiovascular disease, 38
cargo, 37
cell, 9, 38, 39, 47
Centers for Disease Control, 42
centralized, 4, 42
channels, 11
chemical properties, 39
chemotherapy, 38
China, 13, 14, 15, 16, 17, 18, 20, 45
classification, 9
CNTs, 46
Co, 29
Coca-Cola, 21
Coke, 21
collaboration, 7, 12, 23
colleges, 10
commerce, 9
Commerce Department, 29
commercialization, 3, 4, 7, 8, 9, 11, 22, 24, 26, 27, 28, 29, 31, 32, 40, 41, 44
communication, 15, 29, 45
community, 22
competition, 16, 45
competitive advantage, 10, 21
competitiveness, vii, viii, 1, 3, 4, 5, 7, 9, 12, 15, 20, 21, 27, 29, 31, 32, 35, 36, 37, 41, 43, 44, 45, 46
competitor, 11
compliance, 28, 46
components, 4
composition, 9, 19
concentration, 38

confidence, 48
Congress, iv, viii, ix, 3, 4, 5, 6, 7, 10, 21, 27, 29, 31, 32, 35, 36, 37, 40, 41, 42, 49
consensus, 9, 21
construction, 37
consulting, 45
consumers, 12
consumption, 47
contaminants, 39, 40
contamination, 40
control, viii, 5, 29, 35, 37
copper, 39
corporations, 14, 23, 45
corrosion, 40
cosmetics, 5
costs, 24, 28, 46, 49
covering, 20
crack, 40
credit, 26
CRS, viii, 1, 7, 9, 15, 16, 26, 27, 28, 29, 35, 36, 45
currency, 13, 14, 44
customers, 12
cyanide, 40

D

data collection, 32
database, 20
death, 6, 24
decisions, 25, 30, 32
defense, 7, 13, 40
definition, 12, 19
delivery, 38, 47
demand, 23, 24
density, 5, 39
Department of Agriculture, 6, 39, 42
Department of Commerce, 29, 41
Department of Defense, 4, 7, 42, 43
Department of Defense (DOD), 42
Department of Education, 42
Department of Energy, 4, 7, 41
Department of Energy (DOE), 41
Department of Health and Human Services, 42

Department of Homeland Security, 4, 42
Department of Justice, 42
Department of State, 42
Department of the Interior, 42
Department of Transportation, 42
desalination, 39
desert, 24
detection, 6, 38, 39
diffusion, 47
discomfort, 49
diseases, 6
dispersion, 47
displacement, 25
distortions, viii, 2
distribution, 11, 12
diving, 38
DNA, 47
DNA damage, 47
dominance, 11
draft, 6, 39
DuPont, 15, 22

E

economic competitiveness, 32
economic growth, 3, 10, 13, 32, 36
ecosystems, 47
Education, 6, 39, 42
electricity, 39
employment, 7, 43
encapsulated, 38
energy, 6, 13, 38, 47
energy consumption, 47
energy efficiency, 13
engagement, 30, 49
enterprise, 11
entrepreneurs, 11
environment, viii, ix, 2, 3, 6, 8, 13, 22, 28, 32, 36, 46, 47, 48, 49
environmental contamination, 40
environmental protection, 13
Environmental Protection Agency, 4, 6, 7, 28, 41
EPA, 40, 41, 42, 43
epoxy, 40

equipment, 14, 22, 23, 27, 30
equity, 15
EU, 16
Europe, 16
European Union, 16, 17, 18
exchange rate, 13, 14, 44
exchange rates, 13, 14, 44
Executive Order, 31
expenditures, ix, 36, 42
expert, iv
expertise, 23, 24, 28
exploitation, 9
export controls, 29
exposure, 21, 47

F

failure, 23, 25, 47
fears, 48
February, 6, 22, 29
federal government, viii, 2, 3, 4, 7, 8, 21, 23, 25, 26, 27, 28, 29, 30, 32, 43
fibers, 39
film, 39
filters, 16
filtration, 39
financial support, 25
financing, 22
firms, 10, 12, 40, 45
fish, 47
food, 14
Food and Drug Administration, 28, 42
Forest Service, 42
France, 14, 18, 20, 25, 46
fuel, 9, 39
fuel cell, 9, 39
fullerenes, 46
Fullerenes, 47
funding, viii, 2, 3, 4, 7, 13, 14, 21, 22, 23, 24, 25, 26, 27, 28, 30, 32, 37, 40, 42, 43, 44, 49
funds, 3, 4, 12, 30, 42

G

gas, 40
gauge, 9, 44
General Electric, 15
genes, 38
Germany, 13, 14, 18, 19, 20, 44, 46
gifted, 10
global competition, 2
global economy, 32
Global Fund, 44
global leaders, 5, 13, 18, 44
global markets, 29
global trade, 7, 43
globalization, 11
goals, 3, 4, 5, 7, 30, 31
goods and services, 14
government, iv, vii, viii, 1, 2, 3, 4, 6, 7, 8, 12, 21, 22, 23, 24, 25, 26, 27, 28, 29, 30, 31, 32, 43, 48
Government Accountability Office, 31
government intervention, viii, 2, 25
graduate students, 16
grants, 19, 26
greenhouse, 47
greenhouse gas, 47
groups, 42
growth, 3, 10, 13, 14, 20, 22, 24, 32, 36, 38
growth rate, 14
guiding principles, 9

H

harm, 47
health, viii, 2, 3, 7, 11, 13, 22, 27, 28, 30, 35, 36, 37, 41, 46, 48
Health and Human Services, 42
health effects, 46, 48
heat, 38
heavy metal, 39
heavy metals, 39
high risk, 15, 24
high school, 49
high-density memory, 6
high-density**Error! Bookmark not defined.**
memory, 39
high-risk, 26
hip, 5, 13, 18, 44
homeland security, vii, 1, 7, 13, 29, 40
Homeland Security, 4, 42
House, ix, 3, 7, 33, 36, 41
human, 5, 46, 48
humans, 47
hybrid, 27
hydrogen, 38, 40
hydrogen cyanide, 40

I

IBM, 15, 39
id, 7, 8
imaging, 38
impurities, 39
incentives, viii, 2, 10, 22, 24, 26
India, 12, 14
indication, 20
indicators, vii, 2, 5, 7, 9, 12, 32, 43, 44
industrial, viii, 2, 3, 5, 6, 8, 9, 21, 22, 24, 25, 43
industrial policy, 24, 25
industry, 2, 6, 11, 15, 19, 23, 24, 25, 26, 27, 28, 29
infancy, 7, 43
infrastructure, viii, 2, 5, 8, 11, 12, 27, 42, 43
infringement, 27
injury, iv
innovation, viii, ix, 2, 3, 9, 12, 21, 24, 26, 27, 29, 36, 48
Innovation, 8, 10, 26, 42
insight, vii, ix, 2, 36
institutions, 12, 19, 45
instruments, ix, 36, 49
Intel, 15
intellectual property, 11, 20, 21, 22, 27, 41, 44
intellectual property rights, 27
intensity, 49
interactions, 20
international relations, 41
international standards, 30

International Trade, 42
International Trade Commission, 42
inventions, 20, 46
inventors, 19, 27, 45
investment, vii, viii, ix, 1, 8, 9, 11, 12, 13, 15, 24, 25, 26, 28, 29, 31, 36, 42, 43, 45, 46, 48
investors, 12, 14, 23, 24, 25, 31, 45
ions, 23, 29, 31, 39, 47
Ireland, 20
iron, 40
Israel, 13
Italy, 18

J

January, 8, 9
Japan, 13, 14, 15, 18, 20, 44, 45, 46
Japanese, 15, 20
job creation, 10, 13, 32
jobs, 3, 36
jurisdiction, 30
justification, 24

K

killing, 38
Korea, 20, 46

L

labeling, viii, 2
labor, 11, 24, 32, 49
labor force, 32
land, 27
large-scale, 41
lead, vii, ix, 2, 8, 12, 13, 16, 17, 18, 22, 32, 36, 41, 43, 44
leadership, vii, 2, 3, 4, 9, 10, 13, 15, 20, 30, 31, 32, 44, 45
legislation, ix, 33, 36
legislative, ix, 32, 36, 41
Libertarian, 25
licensing, 12
Li-ion batteries, 15

loans, 26
local government, 27
location, 38
long-term, 3, 5, 12, 24
Los Angeles, 39
losses, 39
lung, 46
lungs, 47

M

magnetic, iv, 38
malnutrition, 6
management, 7, 48
manufacturing, viii, 2, 3, 6, 8, 11, 24, 32, 40, 41, 48
market, vii, viii, ix, 1, 2, 5, 7, 8, 9, 10, 22, 23, 24, 25, 28, 31, 36, 43, 46, 47, 48, 49
market share, vii, ix, 1, 2, 7, 9, 31, 36, 43
markets, 2, 11, 25
matrix, 39
measurement, ix, 20, 27, 36, 37, 49
measures, 7, 12, 32, 43, 44
medicine, 47
memory, 5, 6, 39
men, 49
metric, 18, 19
mice, 46
microbes, 47
microfabrication, 23
military, 29, 31
millipede, 39
minorities, 49
misleading, 11, 44
missions, 13, 40
Mitsubishi, 15
modeling, 27
models, 28
molecules, viii, 5, 35, 37, 38, 39, 46
moratorium, viii, 2
multidisciplinary, 42

N

nanocapsules, 40
nanocontainers, 40
nanomaterials, 37, 46
nanometer, 5
nanometers, viii, 5, 35, 37
nanoparticles, 38, 39, 40, 46
nanoscale materials, ix, 27, 36, 40, 41, 46, 47, 48
nanoscience, ix, 13, 36, 48
nanostructured materials, 39
nanotechnology, vii, viii, ix, 1, 2, 3, 4, 5, 6, 7, 8, 9, 10, 11, 12, 13, 14, 15, 16, 17, 18, 20, 21, 22, 23, 26, 27, 28, 29, 30, 31, 32, 35, 36, 37, 38, 39, 40, 41, 42, 43, 44, 45, 46, 47, 48, 49
nanotube, 39
nanotubes, 15, 38, 46, 47
nanowires, 38
NASA, 4, 7, 9, 41, 42, 43
nation, 2, 5, 9, 11, 12, 15, 16, 17, 19, 21, 24, 26, 32, 36, 44, 45
national, vii, 1, 3, 4, 7, 9, 11, 12, 13, 14, 15, 19, 26, 29, 30, 31, 32, 40, 44
National Academy of Sciences, 31
National Aeronautics and Space Administration, 41
National Institute for Occupational Safety and Health, 42
National Institute of Standards and Technology, 4, 7, 26, 41
National Institute of Standards and Technology (NIST), 41
National Institutes of Health, 4, 7, 42, 47
National Research Council, 31
National Science and Technology Council, 5, 6, 13, 39, 41
National Science Foundation, 4, 7, 9, 41
national security, 9, 13
NEC, 15
negative consequences, 48
Netherlands, 18, 19, 20
neurological disease, 38
New York, iii, iv, 25

NIH, 42, 43
NIST, 26, 28, 42, 43
nontoxic, 47
NRC, 31
Nuclear Regulatory Commission, 42
nutrients, 40
nutrition, 6, 39

O

objective criteria, 25
Office of Management and Budget, 4, 42
oil, 39
online, 12
open markets, 30
opposition, 25, 49
optical, 27, 38
optical properties, 38
organic, 39
organizations, 8, 22, 28
OSTP, 42
ownership, 19

P

paper, 15, 16, 17, 18, 33
particles, 40, 46, 47
partnership, 8
partnerships, 27
Patent and Trademark Office, 20, 42, 45, 46
patents, vii, ix, 2, 3, 9, 12, 19, 20, 21, 27, 32, 36, 44, 45
pathogens, 6, 40
perceptions, 49
performance, 37
pharmaceuticals, 47
philosophy, 25
photovoltaic, 38
photovoltaic devices, 38
physics, 17, 18
planning, 24, 27
plastic, 39
play, 21
policymakers, 12

politicians, 25
pollution, 47
polymers, 39
pools, 27
pores, 39
portfolio, 5, 30, 31
positive correlation, 49
power, 6, 13, 38, 44
PPP, 13, 14, 15, 44, 45
preferential treatment, 25
president, 8
President Bush, 4, 28, 29, 31, 42
President Clinton, 3, 22, 37, 41
prevention, 6
prices, 14
private, vii, viii, 1, 2, 3, 5, 8, 9, 11, 12, 15, 22, 23, 24, 25, 26, 27, 32, 35, 40, 43, 45
private investment, 12, 15, 25, 26, 32
private sector, vii, viii, 1, 3, 5, 15, 22, 24, 27, 35, 40
private sector investment, viii, 36
process innovation, 9
producers, 12, 21
production, ix, 9, 10, 11, 12, 13, 15, 19, 23, 24, 26, 28, 36, 38, 44, 49
productivity, 5, 12
profit, 24
program, 9, 10, 25, 26, 30, 33, 41
promote, 37
property, iv, 11, 20, 21, 22, 27, 41, 44
prosperity, 24
protection, 13, 20, 21, 29, 47
protectionism, 24
prototype, 22, 24
prototyping, 27
public, vii, viii, ix, 1, 2, 3, 7, 9, 11, 12, 13, 14, 15, 21, 23, 25, 26, 28, 30, 32, 35, 36, 37, 40, 41, 43, 44, 46, 47, 48, 49
public funding, 14, 40
public health, 13, 48
public investment, vii, viii, 1, 3, 13, 25, 35, 44
public policy, 7
purchasing power, 13, 44
purchasing power parity, 13, 44
purification, 6

Q

quality of life, 3

R

R&D, viii, ix, 2, 3, 4, 5, 7, 8, 12, 13, 14, 15, 21, 22, 23, 24, 25, 26, 27, 28, 30, 35, 36, 37, 38, 40, 41, 42, 44, 45, 46, 48, 49
R&D investments, viii, 13, 14, 15, 25, 28, 35, 37, 41, 44, 45
range, 2, 13, 30
reactivity, 40
recognition, 13, 16, 32
regional, 17
regulation, ix, 5, 28, 36, 41, 49
regulations, 11, 28, 29, 30, 46
relationship, 26
relevance, 16
remediation, 6, 40
repair, 40
research, vii, viii, ix, 1, 2, 3, 5, 7, 9, 10, 11, 12, 13, 14, 21, 23, 24, 25, 26, 27, 28, 29, 31, 32, 35, 36, 37, 38, 40, 42, 44, 46, 48, 49
research and development, vii, viii, 1, 3, 5, 9, 12, 14, 23, 26, 27, 28, 31, 32, 35, 36, 37, 38, 49
Research and Development, ix, 3, 4, 7, 12, 31, 32, 36, 37, 41, 49
researchers, 14, 46
resistance, 30, 37
resource allocation, 30, 32
resources, 5, 10, 11, 13, 25, 28, 30, 31
responsibilities, ix, 3, 7, 36, 37, 41
returns, 15, 23, 24, 26
revolutionary, vii, 1, 6, 31, 37
risk, 15, 24, 48
risk assessment, 48
risk management, 48
risks, ix, 24, 36, 46, 47, 49
Russia, 12, 13, 14
rust, 37

S

safety, viii, ix, 2, 7, 13, 22, 27, 28, 30, 35, 36, 37, 41, 46, 48
salaries, 14
sales, 3, 11
salt, 39
sarin, 40
savings, 39
scaffolding, 10
scalable, 39
scholarships, 29
school, 49
scientific knowledge, 13
scientific understanding, 9, 31
scientists, 10, 11, 12, 15, 29, 47
search, 20, 27
secret, 21
secrets, 21
security, vii, 1, 7, 9, 13, 29, 40
selecting, 16
Self-healing, 40
self-repair, 47
semiconductor, 38
semiconductors, 8, 43
senate, 9
Senate, ix, 3, 7, 8, 29, 33, 36, 41
sensors, 6, 27, 38, 39, 40
services, iv, 2, 14, 24, 27, 44
shape, 22
shipping, 12
short-term, 23
side effects, 38
signals, 24, 25, 38, 40
silica, 38
silver, 47
Singapore, 13
single-wall carbon nanotubes, 46
sites, 6, 38, 40
software, 8, 43
soil, 6, 40
solar, 38
solar panels, 38
solutions, 40
soot, 46

South Korea, 13, 14, 18
space exploration, 7, 40
spectrum, 11, 38
speculation, 6, 37
speed, 27, 32
stakeholders, 27, 48, 50
standards, viii, ix, 2, 22, 27, 30, 32, 36, 37, 49
statistics, 7, 43
statutory, 4, 37
storage, 6, 38, 39
strategic, 3
strategies, 23
strength, vii, 2, 5, 9, 11, 15, 16, 19, 23, 45
strong interaction, 20
students, 10, 12, 16, 29, 30, 32, 44
subjective, 30
subsidies, 24
subsidization, 30
substances, 47
suffering, 6
sugar, 6
sunlight, 38
supply, 10, 12
supply chain, 10, 12
surface area, 27, 39, 46
surgical, 38
systems, 5, 6, 37, 39, 40, 41

T

Taiwan, 12, 13, 14, 17, 18
takeover, 11
talent, 11
targets, 38
tariffs, 30
tax credit, 26
tax credits, 26
tax deduction, 26
taxes, 11
taxpayers, 25
technicians, 29
technological change, 28
technological developments, 25
technology, viii, 2, 3, 5, 6, 8, 9, 10, 13, 19, 22, 23, 25, 26, 27, 32, 35, 37, 41

telephone, 12
therapeutic agents, 47
threat, 24
time, 3, 16, 21, 30, 45
Times Magazine, 8, 23, 48
timing, 48
TIP, 26
tissue, 46
title, 16
Toxic Substances Control Act, 28
toxicity, 46, 47
toxins, 6, 40
trade, 2, 7, 9, 21, 30, 31, 41, 43
training, viii, 2, 30, 41
transfer, 25
translation, viii, 2, 3
transmission, 6, 38
transport, 25
Treasury, 42
TSCA, 28
tumor, 47

U

U.S. Department of Agriculture, 6, 39
U.S. economy, vii, 1
U.S. Geological Survey, 42
UK, 18
uncertainty, 28, 46, 47, 49
United Kingdom, 14, 16, 18, 25
United States, vii, ix, 1, 2, 3, 5, 7, 8, 9, 10, 11, 12, 13, 14, 15, 16, 17, 18, 19, 20, 26, 27, 29, 31, 32, 36, 39, 41, 43, 44, 45
universities, 10, 23, 26, 29
USDA, 28, 42

V

values, 30
variability, 49
vehicles, 27
venture capital, 3, 14, 15, 23, 45
vessels, 12
virus, 39, 40
viruses, 39

W

water, 6, 38, 39, 40
wealth, 3, 21, 25
web, 27
web-based, 50
welfare, viii, 2, 25
well-being, vii, 1
White House, 4, 5, 6, 12, 24, 25, 39, 41
White House Office, 4, 41
wires, 39
women, 49
workers, 12, 24, 30
workforce, 7, 11, 22, 32, 42, 47
working groups, 42
World War, 10
World War I, 10
World War II, 10

Y

yield, 6, 39